Piton

W0246152

Automated Reasoning Series

VOLUME 3

Managing Editor

William Pase, *Odyssey Research Associates, Ottawa, Canada*

Editorial Board

Robert S. Boyer, *University of Texas at Austin*
Deepak Kapur, *State University of New York at Albany*
Hans Jürgen Ohlbach, *Max-Planck-Institut für Informatik*
Lawrence Paulson, *Cambridge University*
Mark Stickel, *SRI International*
Richard Waldinger, *SRI International*
Larry Wos, *Argonne National Laboratory*

Piton
A Mechanically Verified
Assembly-Level Language

by

J STROTHER MOORE

Computational Logic, Inc.,
Austin, Texas, U.S.A.

SPRINGER-SCIENCE+BUSINESS MEDIA, B.V.

A C.I.P. Catalogue record for this book is available from the Library of Congress.

ISBN 978-94-017-3791-3   ISBN 978-0-585-33654-1 (eBook)
DOI 10.1007/978-0-585-33654-1

Printed on acid-free paper

All Rights Reserved

© 1996 Springer Science+Business Media Dordrecht
Originally published by Kluwer Academic Publishers in 1996
Softcover reprint of the hardcover 1st edition 1996

No part of the material protected by this copyright notice may be reproduced or
utilized in any form or by any means, electronic or mechanical,
including photocopying, recording or by any information storage and
retrieval system, without written permission from the copyright owner.

Contents

Preface

Mountaineers use pitons to protect themselves from falls. The lead climber wears a harness to which a rope is tied. As the climber ascends, the rope is paid out by a partner on the ground. As described thus far, the climber receives no protection from the rope or the partner. However, the climber generally carries several spike-like pitons and stops when possible to drive one into a small crack or crevice in the rock face. After climbing just above the piton, the climber clips the rope to the piton, using slings and carabiners. A subsequent fall would result in the climber hanging from the piton—if the piton stays in the rock, the slings and carabiners do not fail, the rope does not break, the partner is holding the rope taut and secure, and the climber had not climbed too high above the piton before falling. The climber's safety clearly depends on all of the components of the system. But the piton is distinguished because it connects the natural to the artificial.

In 1987 I designed an assembly-level language for Warren Hunt's FM8501 verified microprocessor. I wanted the language to be conveniently used as the object code produced by verified compilers. Thus, I envisioned the language as the first software link in a trusted chain from verified hardware to verified applications programs. Thinking of the hardware as the "rock" I named the language "Piton." The trusted chain was actually built and became known as Computational Logic, Inc.'s "short stack."

It is now 1994. The Piton project did not take eight years. Some of what happened in the meantime is relevant and is told as part of the history of the project. But some of the delay is due to my own procrastination. In addition, some thought was given to patenting some of the components of the stack and the publication of some of the Piton results might have compromised that attempt. In the end, we decided it was in the best interests of all concerned simply to publish our results in the normal scientific tradition. I am sorry for the delay.

The Piton project benefited substantially from the contributions of Warren Hunt, Matt Kaufmann, and Bill Young. Warren showed me how to program in FM8502 machine code, helped write the first version of the linker, and produced FM8502 from FM8501 in response to my requests. Matt volunteered to help construct the correctness proof and "contracted" to deliver the proof for one of the three main lemmas. I can think of no higher testimony to his mathematical and clerical skills than merely to point out that he was given a formula involving, at some level, about 500 defined function symbols and two months later delivered his proof—after finding and correcting dozens of bugs. His participation in the proof effort sped the whole project up far more than suggested by his two months of work. Finally, Bill is the first user of Piton—it is the target language of his Micro-Gypsy compiler—and so he has had the burden of being the "bad guy" who always needed some (perfectly reasonable) feature I had omitted. Without him, Piton would be far more of a toy than it is. Bishop Brock helped me when I ported the FM8502 Piton downloader and its proof to the FM9001. I would also like to thank Matt Wilding for his careful reading and constructive criticism of the first draft of this book, his use of Piton to produce a verified NIM program [33], and his energy in getting the first Piton binary images actually downloaded to and running on the fabricated FM9001 device. This actually happened first at the University of Indiana, to which we had sent one of our fabricated devices. Ken Albin helped get the first Piton images running at CLI. Finally, Art Flatau, who wrote and verified the second compiler which produces Piton object code [16], also helped clarify the presentation of Piton in the first draft of this book. Bob Boyer has been very supportive throughout the Piton work, both as a source of technical advice and enthusiasm for the work and its distribution and publication. Mike Smith wrote the infix printer which generated most of the formulas shown here from the Lisp-like s-expression notation actually used by Nqthm. Mike was extremely helpful in producing the final draft of this book. Some of the formulas were "hand prettyprinted" by me and so the responsibility for the typographical errors is on my shoulders not Mike's software. Finally, I would like to thank all my other colleagues at Computational Logic, Inc., and especially Don Good, for making this such a good place for me to work.

Two anonymous referees of the early draft of this book deserve special thanks for their exceptionally detailed and thoughtful comments. The book is much better for their efforts.

This work was supported in part at Computational Logic, Inc., by the Advanced Research Projects Agency, ARPA Orders 6082 and 9151. The views and conclusions contained in this document are those of the author and should not be interpreted as representing the official policies, either expressed or implied, of Computational Logic, Inc., the Advanced Research Projects Agency or the U.S. Government.

As for the name "Piton," I should point out that many climbers eschew their use. They often damage the rock face and, when properly placed, they cannot be easily removed. Because of concern for route protection, continued access to challenging climbs, new technology, and the changing aesthetics of sport climbing, pitons are not often found on the modern climber's rack. They have been replaced by a variety of lightweight removable anchors that come in a plethora of sizes and styles such as nuts, cams, and stoppers. Nevertheless, if I ever fall onto a single artificial anchor, I hope it's a well placed piton.

1

Introduction and History

1.1. What This Book is About

Piton is a simple assembly-level programming language for a microprocessor called the FM9001 described at the machine code level. The correctness of the implementation has been proved by a mechanical theorem prover.

This book is about the exact meaning of the previous paragraph. What is Piton, exactly? What is the FM9001? How is Piton implemented on the FM9001? In what sense is the implementation correct? How is its correctness expressed mathematically? How is it proved? These questions are answered here. Also discussed is the evolutionary character of software, the Piton implementation in particular, and how proof plays a continuing role in its design and improvement.

Should you spend your time reading this book? Don't approach it that way. Read this first chapter and then decide. It won't take long and it informally tells the whole story.

Piton is a simple but non-trivial programming language. It provides execute-only programs, recursive subroutine call and return, stack based parameter passing, local variables, global variables and arrays, a user-visible stack for intermediate results, and seven abstract data types including integers, data addresses, program addresses and subroutine names. Here is part of a Piton program that illustrates the language. The program is printed in an abstract syntax (but the Piton implementation deals with parse trees and does not include a parser). This program is discussed at length later. It is used here merely to suggest the level at which the Piton programmer deals with computation.

```
subroutine big-add (a b n)
        push-constant f        ; push f on the stack
        push-local a           ; push (the value of) a (an address)
loop fetch                     ; pop an address, fetch and push contents
        push-local b           ; push b (an address)
        fetch                  ; pop an address, fetch and push contents
        add-nat-with-carry     ; add the topmost 2 elements of stack
        ...
```

Piton is implemented on the FM9001 via a mathematical function that generates an FM9001 binary machine code image from a given system of Piton programs and data declarations. This function, called the Piton "downloader," is realized by composing a compiler, assembler, and linker. Note that the Piton downloader is not a program running on the FM9001 but a mathematically defined function. It would not be misleading to think of it as a Pure Lisp program that generates FM9001 binary images from Piton programs. Below are the first few few bit vectors in that portion of the image produced from the program above by the downloader.

```
. . .
00001111111000001000100000000001
00001111111000000000010000000010
00001111111000001000100000110011
00001111111000001000100000110011
00001111111000001000100000110011
00001111111000001000110000111111
00000000000000000000000000000000
00001111111000000001000000111111
00000000000000000000000000000000
00000011110000000001000000000010
00001111111000001000110000010100
00001111111000000001000000110011
. . .
```

Of course, a binary image for an undescribed machine is of almost no interest. Perhaps, however, the "core dump" communicates an intuitive appreciation of the transformation wrought by the Piton downloader. This image is essentially the data for an abstract, finite, register-based von Neumann machine, the FM9001, and causes that machine to carry out a certain computation.

It would be interesting to show that the answer delivered by the above Piton program is "the same as" that produced by the FM9001 on the downloaded image. In that case one might say the image is "suitable." The challenge addressed in this book is more general. Roughly speaking, the Piton downloader should produce a suitable binary image for *every* legal Piton program. Of course, this cannot be done because the FM9001 has only a finite amount of memory and Piton programs can be arbitrarily large. But there is a practical sense in which the Piton implementation is correct and it is that sense captured in the theorem proved about it.

The theorem can be stated informally as follows. Suppose p_0 is a "proper Piton state" for a "32-bit wide" "Piton machine." Suppose p_0 is "loadable" onto the FM9001. Let p_n be the result of "running the Piton machine" n steps starting from p_0. Suppose that no "runtime error" occurs and that the final "answer" has "type specification" ts. Then the answer can be alternatively obtained by "downloading" p_0, "running FM9001" some k steps from that initial state, and then interpreting a certain region of memory (given by the "link tables") as representing data of type specification ts. The k in the theorem is constructed from p_0 and n.

Among the interesting technical aspects of the Piton project are that truly abstract objects and operations are implemented on a much lower level processor in a way that is mechanically proved correct, the notion of "erroneous" computation is formalized and exploited to make the compiled code more efficient, and the programmer's "foreknowledge" of the type of the final answer is formalized and exploited to explain how the final binary answer is interpreted. Also interesting is that the Piton correctness theorem implicitly invites the user of the downloader to prove the correctness of the source programs being downloaded. The reason for this is that the theorem applies only to non-erroneous source programs for which the type of the final answer is known. These properties of the source program can only be established via analysis with respect to the high-level semantics of Piton.

1.2. Piton as a Software Project

But these technical aspects are not the main reason the work is interesting. The most interesting aspects of Piton are its reality and its history as a small but representative software project in which mathematical specification and proof play an integral role.

Piton is only part of a much larger body of work demonstrating the current state of the art in proofs about computing systems. The complete body of work is called the "short stack," because it represents a stack of verified components in a simple computing system. Piton is in the middle of the stack. Below it is the FM9001; above it are several compilers that produce Piton object code.

The FM9001 operationally defines a binary machine language. The FM9001 has been implemented at the gate-level by a netlist describing the interconnection of many hardware modules. This netlist has been proved to implement the FM9001 machine language. The verified netlist was mechanically translated into LSI Logic's Netlist Description Language and from that description LSI Logic, Inc., fabricated a microprocessor as a CMOS gate-array. The FM9001 implementation, proof, and fabrication is described in detail in [7]. Above Piton are verified compilers for nontrivial subsets of two high-level programming languages, Gypsy [34] and pure Lisp [16]. These compilers produce Piton object code and have been verified in the same sense that the Piton downloader was verified.

Thus, the short stack provides a means by which a high-level program satisfying certain explicitly stated restrictions can be transformed into a binary image that is guaranteed to compute at the gate-level the same answer computed by the high-level program.

The fact that Piton is in the middle of a fabricated stack increases the credibility of this work as a benchmark. The fabrication of the FM9001 forced upon its designers many complexities and restrictions omitted from earlier "paper" designs. Some of these complexities are visible to the machine code programmer and thus to the Piton downloader. It would have been much easier, but less credible, to design Piton around a machine that provided unlimited resources, stacks, execute-only program space, a variety of data types, subroutine call primitives, etc. Similarly, the fact that Piton is the target language for two compilers forced Piton to provide capabilities that would have been more conveniently omitted from the language.

Another impact of the stack is that its enduring quality has presented Piton with a "maintenance" problem. The version of Piton that is described here is actually the fourth (or fifth if one counts the prototype), each of which was implemented with a downloader which was proved correct. As described in the following history of the project, Piton was implemented and proved correct for one processor, then the Piton language was extended at the request of a user, then the downloader was retargeted to a different host processor, and finally (?) the downloader was extended to allow some user control over the use of memory regions.

It is shortsighted to think of a project producing in one massive sustained effort a "final" piece of software and its correctness proof. Software, even correct software, evolves with the changing needs of the users and hosts. "The" specification and "the" proof evolve too. Every time Piton was changed it was "reverified." Technically of course a new theorem was proved about a new collection of mathematical functions. But the basic structure of the previous proof and the previously developed library of lemmas were both reused.

Piton is the first example of stacking mechanically verified components of significant size. While Piton is relatively simple it is a representative software project. To my knowledge, it is the most complex compiler/linker yet verified mechanically. More significantly, this piece of software has evolved in a realistic environment and its proof has evolved also.

1.3. About This Book

From one perspective, this book can be summarized as presenting two computational paradigms, a compiler that implements one in terms of the other, a very precise statement of its correctness, and a brief description of a proof of that statement. From that perspective, the book is fairly conventional.

But the book is quite unconventional by virtue of the fact that all of the foregoing is ultimately couched in a mathematical formalism. That is, the semantics of both Piton and the binary machine are presented as systems of mathematical equations describing the operations of the two abstract machines. The downloader is presented as a recursively defined function that maps a Piton state to an FM9001 state. The statement of correctness is a mathematical formula that can be derived as a theorem from the foregoing equations, using the most primitive deductive steps such as replacements of equals by equals and mathematical induction.

The formal logic is explained informally before much use is made of it. Everything else is explained informally as well. Nevertheless, the mathematical formalization is offered as the definitive expression of the specification. Thus, this is really two intertwined books, one written in English and the other written in the formal logic. It is hoped that this makes the book clearer and more accessible than it would otherwise be. If you are uncomfortable with formalism, skip those parts. If you find the informal remarks confusing or ambiguous, read the formal parts with the assurance that they are complete and unambiguous.

For whom is this book written? What do you have to know to understand it? What will you gain if you spend the time to read it?

The answers depend, in part, upon which of the two books is considered. But the first prerequisite of both books is an open mind with respect to the question of what mathematics can bring to the production of reliable computing systems. No theorem can be proved about a physical device, such as the chip of silicon fabricated from the netlist for FM9001. Nothing can guarantee that your "verified chip" will work as specified the next time you use it. Subtle chip torturing schemes should come immediately to mind. Subject your trusted chip to large static discharges and see how it behaves then. Bombard it with cosmic rays. Let time and metal migration ravage its pathways. There are no guarantees in this world.

These observations are so obvious that the practitioners of mathematical methods often fail to make them and then find their work attacked on the grounds that "false" guarantees are made. If you are inclined towards that view, this book is not for you. I here embrace mathematical methods with the realistic expectation and understanding of what they can do and what they cannot do. Theorems are proved about mathematically defined objects, not physical ones. These mathematical objects might constitute models of physical objects. The netlist for the FM9001 is one model of the fabricated chip. Another model is the FM9001 machine code interpreter. It is a theorem that those two models are equivalent, and the fabricated chip is more or less directly constructed from the former model. Lower level models, dealing with layout in 3-space, voltages, timing, etc, could be produced and might be useful. But no matter how far down the lowest model is pushed, there is still an enormous and unbridgeable gap between what our theorems talk about and that piece of silicon. The same statement can be made about the use of mathematics in the construction of larger-scale engineering artifacts. No theorem can guarantee that a given beam will support a given load. Nevertheless, mathematics is useful in engineering.

In my personal experience, when a software system has failed to perform according to the implementor's expectation the problem has most often been one that could have been prevented by specification and proof. That is, the underlying physical devices were apparently performing in conformance with their abstract mathematical specifications but the applications programs were logically erroneous in ways that could be detected by careful symbolic analysis. One might question the *cost* of establishing "logical correctness"—how much symbolic analysis is necessary, what level of training is necessary to do it, and how long it might take—but the *value* of establishing logical correctness is here taken for granted.

The informal book has been written for the computer scientist or computer science student. Knowledge of fundamental programming concepts is taken for granted. Thus, you should be familiar with such concepts as registers, memory, program counters, addresses, push down stacks, arrays, trees, lists, atomic symbols, jumps, conditionals, and subroutine call. In addition, I assume familiarity with the elementary mathematical concept of function. Even in the informal parts, I assume you are willing to deal with some formal notation, namely that for function application, including expressions built up from nested function applications. If, for example, 'f' is a function of two arguments and 'g' is a function of one argument, then you are expected to understand what is traditionally written as 'f$(x, g(y))$.' In a context in which the variables x and y have some understood values, that expression denotes the value of the function 'f' when it is applied to (i) the value of x and (ii) the value of the function 'g' applied to the value of y. Finally, it would be helpful if you are comfortable with the notion of recursively defined mathematical functions, i.e., functions "defined in terms of themselves."

As for the formal book contained herein, virtually no computer science background is required to understand it. After all, the main theorem has been proved by a machine! Thus, except for the logic (which is only informally explained here), everything you need to understand Piton, the FM9001, the downloader, and the theorem is explicitly presented in complete detail. Nothing is taken for granted beyond the ability to read the formulas in the logic (and a good memory for details).

The "logic" used here, called the Nqthm (or Boyer-Moore) logic, is technically a "first order theory" constructed from a first-order, quantifier-free logic of recursive functions and induction by adding axioms describing the Booleans, integers, ordered pairs, and atomic symbols. Readers familiar with formal mathematical logics in general will recognize this one as exceedingly weak and simple. The logic is explained in detail in [5]. If you don't already know the logic and must learn it from the informal description here, it would help if you were comfortable with the basic idea of formal mathematical logic, say as presented in [32]. In addition, it would also help if you knew the programming language Lisp because the logic can be viewed as a simple dialect of pure Lisp.

Perhaps the most important role of this book is to document the state of the art in mechanized verification of computing systems, circa 1990.

I hope for more, however. Mathematical methods have a contribution to make in the production and maintenance of reliable hardware and software. By investing your time to read this book you will come to understand better the problems and promises of those methods.

One reason this work is a useful benchmark in the progress toward the use of mathematical methods is that the proofs have been mechanically checked. This insures that all assumptions are written down. It is conceivable (but very unlikely) that some of the explicit assumptions are unnecessary. And everything could be said in other ways. Nevertheless, this work represents an upper bound on what must be said to nail down the correctness of a realistic compiler and linker. The problem is no worse than this. Furthermore, it is possible to say everything so precisely that a machine can follow the argument and check it for you. Finally, it is possible to maintain the proof as the software changes so that you can rest assured that your patches and modifications are right.

1.4. Mechanized Mathematics and the Social Process

The role of mechanical theorem proving is not emphasized in this book. Yet in a certain sense it permeates the discussion. Were it not for the use of the theorem prover, the use of a formal mathematical logic would not be necessary. The usual mix of formal notation and the precise English found in traditional mathematical textbooks would suffice. Formal notation was used so that the proofs could be checked mechanically. Why was traditional (informal) mathematics rejected? The reason is that traditional mathematics crucially depends upon the so-called "social process" to vet newly minted proofs. This dependence is there for good reason:

people make mistakes. Until a proof is carefully scrutinized by a large number of authorities in the field who are personally intrigued or interested in the result, it is rightly regarded as more of challenge than a reassurance. It is only after proofs have been vetted that the reader can feel comfort and security in the presence of a formula labeled THEOREM. In mathematics this process often takes years and sometimes takes decades.

A major difficulty with the application of mathematical methods to computing systems is that most theorems about computing systems are inappropriate for the traditional social process. That is true of the Piton correctness theorem in particular and it serves as an illustration of the general difficulty. In the first place, the Piton theorem is simply too large—this entire book is more or less devoted to its accurate statement. In the second place, it is of personal interest to too few authorities. In the third place, it changes periodically as Piton evolves. The publication of an informal proof of the correctness of the Piton downloader would be a waste of paper. Until vetted, it would neither establish the correctness of the downloader nor serve as a reliable indicator of how much effort is necessary to do that. In the meantime, projects assuming the correctness of the Piton downloader would be of questionable integrity—and the "meantime" might be quite long.

Mechanical theorem proving offers an escape from this dilemma. Instead of subjecting the Piton downloader to the social process, Piton's correctness is formally stated and proved mechanically with a theorem prover that has been subjected to the social process. Mechanical theorem provers are natural recipients of the scrutiny of the social process. Their soundness is a clearly posed proposition that is understandable to virtually all mathematically trained people. Their soundness is of primary concern both to their developers and their best users. But their generality and increasingly successful application draw the attention of a wide field of authorities.

The mechanical theorem prover used in this work, Nqthm, has survived the scrutiny of that community now for almost two decades. Granted, several different versions of the system have been released during those 20 years and, as with any piece of software, all bets are off as soon as any change is made. But only one soundness mistake has ever been found in a released version of Nqthm and the system has been extremely visible and widely used. In fact, because of the mathematical nature of many of the theorems proved with Nqthm (including Gödel's incompleteness theorem [31], Gauss' law of quadratic reciprocity [30], and the Paris-Harrington extension of the finite Ramsey theorem [22]) it is possible it has received more than its share of scrutiny. While this does not establish the soundness of Nqthm, it increases the confidence in Nqthm to the point where I do not consider Nqthm to be the weak link in the chain establishing the correctness of the Piton downloader.

The weak link is in the statement of the theorem proved. It is hard to imagine being more certain that the Piton correctness formula is a theorem.[1] But can the

---

[1]Hard, but not impossible. More certainty could be gained by having the theorem prover create a formal proof object which is then checked by a simpler piece of code which has survived the social process.

formula be characterized as saying "the Piton downloader is correct?" Note that the formula does not say, literally, "the downloader is correct." Whatever it says, it takes roughly a book to write down! The social process can profitably be applied to that formula as a formal expression of the intuitively understood notion of correctness. This is an enterprise that finds wider appeal than the correctness of the proof itself largely because the issues raised are of more general interest than how to compile Piton for the FM9001.

It bears pointing out, however, that the appropriateness of the informal interpretation of the Piton theorem is of no consequence to the users of the short stack. The formula that states "the stack is correct" says that certain high-level computations can be equivalently carried out by a gate-level netlist operating on a binary image produced by the composition of certain transformations. The formula does not mention Piton and the user of the stack need not know about Piton nor trust its implementation. The formula stating "the stack is correct" is proved using the Piton theorem as a lemma. In particular, the stack's correctness relies on the *formula* proved about Piton, not on any informal characterization of it. Whatever the formula says, it was adequate to allow the stack proof. That is the beauty of formal proof: vast complexity can be swept away by it.

1.5. The History of the Piton Project

To make real the evolutionary character of Piton, we must tell its story. In 1982, a graduate student in our "Programming Languages" class at the University of Texas, presented his class project to Bob Boyer and me. The student was Warren Hunt and his project was the Nqthm formalization of part of the Z80 microprocessor. He was frustrated by his inability to specify the processor more fully because the available documentation was incomplete and ambiguous. So he undertook the specification, implementation, and proof of a microprocessor of his own design and Boyer and I agreed to be his supervisors.

Hunt named his processor FM8501. The FM8501 is a 16-bit, 8 register general purpose processor implementing a machine language with a conventional orthogonal instruction set. At the highest level of his specification, Hunt described his machine with a mathematically defined function that is most easily described as an interpreter for the machine code language. He defined the function in the Nqthm logic. Hunt also formally described the combinational logic and a register-transfer model that he claimed implemented the instruction set. The Nqthm theorem prover was then used to prove that the register-transfer model correctly implemented the machine code interpreter. This work is described in Hunt's PhD dissertation, [18].

The FM8501 was never fabricated; it exists only as a "paper machine." In a sense, its specification style made it impossible to fabricate by conventional means. The Boolean primitives rather than standard hardware components were used to describe the combinational logic, interconnection was implicitly represented by function application, "fan out" was suggested by replication of expressions, etc. Producing a verifiable register transfer model that could also be used more or less directly with conventional CAD tools to fabricate the processor would inspire much of Hunt's subsequent hardware verification work.

But the existence of a verified design for a general-purpose processor clearly suggested the idea of building a verified processor and using it as the delivery vehicle for some "trusted system" such as an encryption box, embedded software, a verified high-level language, or perhaps even a program verification system. Unless one builds such tools in machine language, it would be necessary to implement higher level languages on the processor. To maintain the credibility of the final system, the implementation of those languages should be verified all the way down to the machine code provided by the processor. While we knew the FM8501 would not be built, we assumed that the problems of implementing verified higher level languages could be explored with the FM8501 in the expectation that the solutions could be carried over to the verified processor that would be eventually fabricated.

In September, 1986, Warren Hunt and I sketched a stack based assembly-level language for FM8501 and implemented in Nqthm an assembler and linker for a small subset of it containing about 10 instructions. This was done without defining the formal semantics of the language; we viewed the assembler merely as a convenient way to produce machine code. Properties of the machine code programs could be proved directly from the FM8501 definition. This view of an assembly-level language was exactly that taken contemporaneously by Bevier in [2]. It has since been carried quite far by Yu, who has mechanically proved with Nqthm the correctness of 21 of the 22 Berkeley C string library programs by reasoning about the binary machine code produced by the GCC compiler for the Motorola 68020 [6, 35]. However, one problem with this approach is that the machine code programs thus produced can, in principle, overwrite themselves during execution. This complexity must be dealt with when proving the programs correct and generally requires hypotheses about where in memory each program is located and where data resides.

The desire to prove theorems about our programs at a higher level than the FM8501 definition forced us to define the semantics of the "assembly language" formally. We decided to make our "assembly language" programs "execute only" so they could be treated as static objects during proof. In addition, to make it easy to compile higher level languages into the "assembly language" we decided that it should provide the abstractions of stacks, local variables, and subroutine call and return. Thus, the "assembly language" we designed for FM8501 is an unconventional one for a machine like the FM8501 because it provided abstractions not directly supported by the machine.

Being unfamiliar with the proof-related problems of providing such abstractions we decided, wisely, to explore the problem in a feasibility study. To that end, we designed a "toy language" that contained only four instructions: a simplified **call** (with no provision for formal parameters), **return**, a variable-to-variable **move**, and an increment-by-2 instruction, **add2**. This language was called 'h' (for *h*igh level). We defined a 10 instruction low-level machine, called 'l' which was a simplified FM8501, to which it was just possible to compile 'h'. We implemented 'h' via a compiler and link-assembler and formulated what was meant by the correctness of the implementation.

Hunt then turned his attention to the problem of how to specify and verify a microprocessor in a style that would support fabrication by conventional means. I proceeded to prove the correctness of the implementation of 'h' on 'l'. By this time, we had also both left the University of Texas and begun working at Computational Logic, Inc. (CLI).

The "toy proof" was completed by September, 1987, a year after Hunt and I began work on the language design. During the first seven months of that year, the project was staffed by 2 men working roughly 8 hours per week. During the last 5 months, the project was staffed by 1 man working roughly 8 hours a day, less about one month of time off. Thus, 7 man-months were devoted to the Piton feasibility study. The importance of this early phase of the project cannot be overemphasized. In the first proof attempt I failed to disentangle several issues. As a result I needed an inductively provable theorem that could not be stated without the invention of some abstractions that were more general than any I had used in the implementation. But these abstractions, once made explicit, could be used in the implementation to make it simpler and more modular. (See the discussion of the "hidden resource problem" on page 149.) Had I encountered the problems for the first time in the vastly more complicated setting of Piton and FM8501, rather than 'h' and 'l', their solution would have been much more costly.

Work on the full-blown language, implementation and proof began in September, 1987. The name "Piton" was chosen, for the reasons described in the Preface. When the Piton project began, the intended hardware base was FM8501. Early in the project I requested two changes to FM8501, which were implemented and verified by Hunt. The modified machine was called the FM8502. The changes were (a) an increase in the word width from 16 to 32 bits and (b) the allocation of additional bits in the instruction word format to permit individual control over which of the ALU condition code flags were stored. Because of the nature of the original FM8501 design, and the specification style, these changes were easy to make and to verify. Indeed, the FM8502 proof was produced from the FM8501 script with minimal human assistance.

After several months of working on the project, I had defined Piton, the implementation, the concepts used in the correctness theorem, and the abstractions necessary for the proof. I had also stated the main theorem and stated the three main lemmas that would be needed to prove it. Each of these three lemmas represents a commutative diagram in a hierarchical decomposition of the problem. The general character of the decomposition and the issues dealt with in each of the layers were discovered in the 'h' to 'l' feasibility study. Having clearly specified the proof problem I enlisted the assistance of Matt Kaufmann, another colleague at CLI. Kaufmann undertook to prove one of the main lemmas, using his interactive proof checker for the Nqthm logic, while I worked on the other two, using the Nqthm theorem prover.

Our proofs proceeded in parallel. During the course of the proof many "bugs" were discovered. These bugs sometimes rippled out to other layers of the main proof, since the functions in our separate problems were not disjoint. For example, when Kaufmann found and repaired a bug in one of "his" functions it might require me to change "my" copy of that function and related proofs. We therefore kept in close communication about our progress even though we worked independently. When changes were necessary, we discussed them and each of us would informally investigate how the proposed changes would affect his work. If the change impacted the other person's proof, the person managing that proof would wait until he got to a nice stopping place, make the change, reconstruct his proof to that point, and then continue. This was much less expensive (in terms of our "context switching") than it would have been had the proofs been constructed sequentially because all three of the main lemmas put different constraints on the functions. The successful proof of one of the lemmas did not necessarily mean all of the functions involved were "right" so iteration and collaboration were important.

By May, 1988 the entire proof had been mechanically checked, but some proofs had been done with the Nqthm theorem prover and others had been done by Kaufmann's proof checker (according to which of us managed the task). The sloppy iteration described above had resulted in the proof having a patchwork appearance that would have made its "maintenance" (i.e., later extension to new language features) difficult. Therefore, I spent three weeks cleaning it up and in the process converted Kaufmann's proofs to Nqthm scripts.

Meanwhile, another CLI colleague, Bill Young, was implementing and verifying a compiler from a subset of the Gypsy language to Piton. Young started with the version of Piton I had defined back in the fall of 1987 and had modified it occasionally to meet the needs of his project. He had not tracked our changes nor we his. In May, after "completing" the Piton proof, Young and I agreed upon the "new" Piton, which was essentially my version with seven new instructions added for his compiler. Technically, every abstract machine in the Piton proof except for FM8502 had to be altered and all of the proofs redone. However, this was done in less than a week because of the way Nqthm had been used and the way the main proof had been decomposed. The key aspect of Nqthm is that the user does not give it specific proofs to check but rather "teaches" it how to prove theorems in a given domain. This "teaching," which might more appropriately be called "rule-based programming," is done by giving it a set of lemmas to prove, lemmas which are subsequently used as rules to guide Nqthm's search. Because I had built a general purpose set of rules for each layer in the proof, minor modifications to theorems in a layer could be accommodated by the proof strategies I had programmed. The new instructions were added by choosing a similar old instruction and visiting every occurrence of that old name in the proof event files. For each formula involving the old instruction an analogous formula involving the new instruction was inserted. With minor exceptions the resulting transcripts were automatically processed. When the automatic processing failed it was because of some relatively deep problem specific to the new instruction (e.g., that integer less than can be computed by taking the exclusive-or of the negative and overflow flags after a subtraction).

Thus, at the end of May, 1988, the "new" Piton was completely implemented and verified. The work was described in a CLI technical report [26] in June, 1988.

The completion of CLI's short stack was marked by the final "Q.E.D." in Young's compiler proof, which also occurred in 1988. The stack consists of Hunt's FM8502, Piton, Young's Micro-Gypsy compiler, and Bill Bevier's KIT operating system, each of which was described in dissertation-length reports [18, 26, 34, 2] and a special issue of the *Journal of Automated Reasoning* [3].[2]

However, the stack as described in 1988 was only a theoretical exercise, because the FM8502 could not be fabricated. But Hunt, working in close collaboration with Bishop Brock of CLI, had continued to make progress towards a formal hardware description language expressed in Nqthm and the use of that language to implement a microprocessor quite similar to FM8502. The design and proof of that processor, called FM9001 [7], was completed in June, 1991. The verified netlist, the necessary test vectors and signal pad assignments were delivered to LSI Logic, Inc., in July, 1991. A fabricated FM9001 was returned to CLI about six weeks later. The need to port Piton to the FM9001 was obvious.

FM9001 differs from FM8502 primarily in that instructions are in a different format and the instruction set is slightly different. However, when Hunt and Brock developed FM9001 from FM8502 they explicitly considered Piton's use of the FM8502 instruction set. For each FM8502 instruction-instance used by the Piton compiler, they included an FM9001 instruction that could provide the same functionality. However, porting the proof from FM8502 to FM9001 was slightly harder than suggested by the instruction set changes alone. FM9001 uses a different formal representation of memory—a binary tree instead of a linear list—and uses a different formalization of bit vectors—lists of Booleans instead of the 'bitv' shell.

In May, 1991, just before the FM9001 proof was completed, I borrowed Brock's Nqthm library for FM9001 and ported the lowest of Piton's three commutative diagrams to it. In that diagram, the upper level abstract machine was like FM8502 but provided symbolic machine code which the downloader's "code linkers" converted to absolute binary. I implemented that abstract machine on the FM9001 by changing the code linkers. For convenience I did it in two steps, introducing a new intermediate machine that was the FM9001 but with a linear memory instead of a tree structured one. Thus, the final Piton proof has four layers, not three. Having convinced myself, in roughly a week's worth of work, that it would be easy to port Piton to the FM9001, I put aside the project and waited until the FM9001 proof was complete and the device was fabricated. In October, 1991 I completed the initial port to FM9001. It took two weeks because of a technical problem discovered: a certain invariant had to be proved of each of the intermediate machines in order to relieve an assumption made when introducing the new memory model. (See the discussion of the "plausibility assumption" on page 159.) This invariant could have been proved in the earlier work but was not necessary.

But the intention of using the Piton downloader to generate binary images for the fabricated device imposed some additional requirements. For example, the FM8502 downloader generated a binary image in which the memory was loaded with compiled code starting at memory address 0. But when the fabricated FM9001 was sitting in its test jig at CLI it would be necessary to have certain debugging and i/o

---

[2]This book is not about the short stack *per se* but we would be remiss if we did not cite some of the related work. The interested reader should see the bibliographic citations on hardware verification and operating system verification in the papers cited above as well as [7] and the work reported in [14] and [20].

code in the low part of memory. Thus, the downloader had to be changed so that (i) Piton's "data segment" was loaded at a specified address in memory, (ii) compiled code was shifted so that it was above the data segment not below it as in the FM8502, and (iii) the low part of memory was loaded with user-specified "boot code." The correctness theorem and its proof had to be changed again. This work took a few days and was done in November, 1991.

At about the same time, Matt Wilding of CLI began to think about writing a verified application program in Piton and running it on the fabricated device. He chose to implement the puzzle-game Nim. Traditionally, the game is played with piles of stones. The two players alternately remove at least one stone from exactly one pile. The player who removes the final stone loses. Wilding implemented in Piton an algorithm which plays for one side. His program chooses its move via an algorithm which involves the bit-wise exclusive-or of the number of stones in each pile. He proves that his strategy wins when a win for its side is possible. This work is reported in [33]. The 300-line Nim playing program was compiled and run on an FM9001 in April, 1992.

For practical reasons, Wilding made several more changes in the Piton downloader. For example, if the compiled code is to be loaded high in the FM9001 memory, say at address 2^{30}, it is impractical for the downloader to construct the initial FM9001 memory. Therefore, Wilding modified the verified downloader so as to return the "relevant" part of the image. This indicates that I should have packaged the downloader differently and proved a different main theorem. I have not yet formulated the "new" theorem because I have been busy with other projects and need more experience with how Piton is actually used in connection with the fabricated device.

Our purpose in giving this rather long and involved history is two-fold. First, it indicates the amount of work involved in the Piton proof and how long it took. Second, it indicates the evolutionary nature of the Piton project. Piton would be much less convincing as a demonstration of the use of mathematical methods in software development had it sprung, complete and correct, from the mind of a single person working in isolation and then simply remained static in all of its verified correctness. Instead, as with most software, it evolved and its statement of correctness and the proof of that statement evolved with it in response to pressures from both above and below.

Having given the history, however, we will describe Piton as it now stands simply to keep the presentation as clear as possible.

1.6. Related Work

The compiler correctness problem was first formally addressed by McCarthy and Painter in 1967 [24]. Their proof was done by hand in the traditional mathematical style and concerned the correctness of a compiler for arithmetic expressions. Within the next five years, several more compiler proofs were published, all done without mechanized assistance, and mainly devoted to the problem of language semantics and convenient logical settings for program proofs. Among the most important papers are those by Burstall [8], Burstall and Landin [9], London [23], and Morris [27]. In 1972, Milner and Weyhrauch [25] published a machine checked proof of a compiler somewhat more ambitious than the expression compiler of McCarthy and Painter. Successive mechanically checked proofs of simple compilers were also described by [10], [1], [5] and [12].

A significant milestone in mechanized proofs of compilers was laid down by Polak [29] in 1981; his work is important because the compiler verified is actually a program in Pascal which dealt with the problem of parsing as well as with code generation. We contrast Polak's compiler to the Piton downloader which can be described as a mathematically defined compilation algorithm operating on abstract syntax. The distinction between a correct compilation algorithm and a correct compiler implementation was apparently first made explicit in [11]. It suggests the idea of decomposing the proof of a practical compiler into two main steps, of which the Piton proof is one. The remaining step is to show that a particular program generates the binary image described by our mathematical function. However, there is another approach which we discuss below.

Historically speaking, our Piton work and Young's Micro-Gypsy [34] are the next major milestones in compiler verification. We have already discussed them.

Shortly after the Piton work was completed, Joyce [19] reported what is probably the compiler verification effort most similar to Piton. He mechanically verified a compiler for a very simple imperative programming language using HOL [17]. Unlike Piton, Joyce's language is not intended for use and only supports simple arithmetic expressions (indeed, it is limited to +-expressions in which one argument is a variable symbol), variable assignments, and a **while** statement. The target machine for the compiler is Tamarack, a verified machine simpler than but fundamentally similar to the FM9001. In particular, the Tamarack memory is finite and contains only natural numbers representable in a fixed number of bits. Tamarack has also been fabricated.

Joyce models the semantics of his machines denotationally. However, the execution of a program is modeled by a sequence of states, where a state is a function from variable names to values. This is not unlike our operational semantics. It is not clear that denotational semantics provides much leverage at this level in a verified stack. The vast majority of our proof concerned aspects of our formalization that would not be affected by the substitution of denotational semantics or higher-order logic for our simple operational semantics and first-order logic. That is, most of our lemmas established first-order facts about arithmetic, bit vectors, tables, sequences, trees, and iterative or recursive algorithms that manipulate such finite, discrete inductive data structures. We suspect that Joyce's proof could be similarly characterized and that the first-order fragment of his proof would be similarly large had his source and target machines been as elaborate and complicated as our own. Furthermore, many of our state-level proofs have direct analogues in denotational proofs. Joyce speculates however that our operational approach might make more difficult the eventual verification of high-level programming languages and application programs in those languages. That may well be the case but remains to be seen. We offer in evidence to the contrary the somewhat remotely-connected fact that the Nqthm logic has sufficed for the statement and mechanically checked proof of such deep results as

Gödel's incompleteness theorem and the Church-Rosser theorem of the lambda calculus (see [31]), both of which involve abstract, high level formal systems, *albeit* not conventional programming languages.

The difference of our semantic approaches notwithstanding, the fact that Joyce's work targets a realistically limited machine makes his work quite similar to ours.

The work of Curzon [15], which deals with an assembly-level language for an abstract version of the VIPER microprocessor (see [13]), is also similar to our work in that some of the resource limitations of a realistic host machine are confronted and the proof is machine checked with HOL. However, Curzon's compiler targets an object language that is symbolic and which has an infinite address space; the problems of generating absolute addresses and "linking," dealt with in our Piton work, is not considered.

Significant additional research into compiler verification includes a "hand proof" by Oliva and Wand [28] for a subset of Scheme compiled to a byte-coded abstract machine (which was in fact implemented more or less directly) and the work of Bowen and the ProCos project [4].

In 1992, Flatau [16] described a mechanically checked correctness proof for a compiler from a subset of the Nqthm logic to Piton. Given that our Piton downloader is a function defined in the Nqthm logic the obvious question is "can you compile the Piton downloader with Flatau's compiler?" Unfortunately, the answer is "not yet." The subset handled by his compiler does not include Nqthm's user-defined data type facility, which is used by Piton. This would not be hard to change, since list structures could be used in place of user defined types. The subset does include user-defined recursive functions and dynamic storage allocation (e.g., 'cons'), which are the key ingredients. Thus it is not hard to imagine progressing to the point at which the Piton downloader can be compiled to Piton and thence to the FM9001 via the one-time execution of Flatau's compiler and our downloader. Thus, we can imagine mechanically converting the downloader from a "mathematical compilation algorithm" to a "verified compiler implementation" running on the FM9001, capable of producing suitable FM9001 binary images from Piton systems. As for the syntax question, a parser would still be needed, in the form of some implementation of Lisp's **read** (since Piton is actually written in s-expression form) and suitable input/output primitives. We do not further pursue such "dreams" in this book, except to note that the stack offers wonderful opportunities for advancing the state of the practice.

1.7. Outline of the Presentation

In Chapter 2 we informally present the Nqthm logic.

In Chapter 3 we informally describe the Piton programming language. We essentially adopt the style of a conventional primer for a programming language. We discuss such basic design issues as procedure call, errors, the various resources available, etc. We exhibit many examples. We summarily describe selected instructions. In Appendix I we describe each of the 65 Piton instructions in this informal manner. The material in Chapter 3 and Appendix I is spiritually correct but often incomplete.

In Chapter 4 we illustrate Piton and the ideas discussed in Chapter 3 with a thoroughly worked example. In particular, we deal with the problem of "big number addition." We explain (both informally and formally) what "big numbers" are, how to "add" them, and what the relation is between addition and big number addition. We then exhibit a Piton program that purportedly does big number addition. We exhibit a Piton initial state in which a particular big number addition computation is set up and we show the state obtained by running that initial state. We then exhibit the formal specification of the Piton program, we comment on the utility of our style of specification, and we discuss the mechanically checked proof that the program satisfies its specification. We return to this example when we discuss the implementation of Piton on FM9001 and the correctness theorem for the implementation.

In Chapter 5 we briefly sketch FM9001.

In Chapter 6 we state the correctness theorem for the FM9001 implementation of Piton, we informally characterize the various predicates and functions used in the theorem, and we explain the intended interpretation of the theorem. We then illustrate how the correctness theorem can be applied to the big number addition program developed in Chapter 4.

In Chapter 7 we explain how Piton is implemented on FM9001. We give an example of an FM9001 core image produced from a Piton state, we explain the basic use of the FM9001 resources in our implementation, and we then discuss each of the phases of the implementation: resource allocation, compilation, link-assembling, and image construction.

In Chapter 8 we discuss the proof of the correctness theorem. Since our primary motivation in this book is to convey accurately what has been proved we do not give the entire proof script. The script may be obtained electronically by following the directions in the /pub/piton/README on Internet host ftp.cli.com.

The book contains five appendices. Appendix I summarizes the Piton instructions. Appendix II contains the equations that define Piton. Appendix III contains the equations that define the machine language of FM9001. Appendix IV contains the equations that define the implementation of Piton on FM9001. Appendix V contains the statement of the correctness result and the definitions of all of the concepts used in that theorem (except those contained in the foregoing appendices). Each of these formal appendices begins with a brief, informal "guided tour" through the system defined. With the exception of these guided tours, the material in these four appendices is completely formal and self-contained (given the Nqthm logic as described in [5]). The correctness theorem requires this much material *simply to state*. The *proof* of the correctness theorem requires the formal definition of hundreds of additional functions to characterize the semantics of the intermediate abstract machines implicitly used by our compiler's internal forms. Of course, all of these concepts are listed in the electronically distributed Nqthm proof script.

This book is exhaustively indexed. Approximately 600 function names are defined in Appendices II-V. The index indicates the page number on which each function symbol is defined and lists the page numbers in the formal appendices on which each function is used.

2

The Nqthm Logic

In order to specify Piton formally we need a formally defined language. The language must permit us to discuss such concepts as n-tuples, stacks, variable symbols, assignments or bindings, etc. With such a language we could define the semantics of Piton operationally. Such a definition would take the form of a function 'p' taking two arguments, a formal Piton state, s, and a natural number, n, such that $p(s, n)$ is the result of running the Piton machine n steps starting from s.

We would like the definition of 'p' to be *executable* so that if we had a concrete starting state s and some particular number of steps n, we could evaluate $p(s, n)$ to obtain a concrete "final" state. That is, the language in which 'p' is defined is some sort of programming language.

Finally, we would like to reason about 'p' and other functions defined within the language. Thus there must be some notion of truth in the language and a means of deducing "new" truths from "old" ones. To put it another way, we seek a formal logical theory. Such a theory is composed of a formally specified syntax defining a language of formulas, a set of axioms, and some rules of inference. Intuitively, the axioms are just formulas that are taken to be valid ("always true") and the rules of inference are validity preserving transformations on formulas. A "theorem" is any formula derived from the axioms or other theorems by the application of a rule of inference. Obviously, every theorem is valid. A "proof" of a formula is just the derivation of the formula as a theorem. A proof of a formula thus demonstrates that the formula is valid and so a formal logical theory provides a means of determining some truths. It is possible to build a machine that can check that an alleged "proof" is indeed a proof: just check that every rule cited in the derivation is one of the rules of inference, that every alleged axiom is one of the axioms, and that each reported application of a rule of inference actually yields the formula reported. Such a machine is called a "proof checker." It is even possible to build a machine that searches through all the possible proofs to try to find a proof of a given formula. Such a machine is called a "theorem prover."

Formal logical theories are a dime a dozen. Executable formal theories are somewhat less common. Executable formal theories that are supported by mechanical proof aids are still more rare. Among them is the so-called "Boyer-Moore" or "Nqthm" logic. That is the one used in this work.

The Nqthm logical theory is described in detail in [5], where its mechanical theorem prover is also described. Roughly speaking, the theory can be obtained from first-order predicate calculus with equality by restricting one's attention to quantifier free formulas, adding axioms to define primitive functions for dealing with two Boolean objects, the natural numbers, the negative integers, symbols, and ordered pairs, adding mathematical induction as a rule of inference, and permitting extension by the addition of terminating recursive definitional equations and a schema for adding new inductively constructed data types. With the right syntax, the logical theory just described is first-order pure Lisp. By "first-order" here is meant that functional parameters are disallowed. A complete description of the Nqthm logic is beyond the scope of this book. The interested reader should see [5]. In this chapter the logic is sketched as though it were simply a programming language—but the reader is urged to remember that underlying it are the axioms and rules of inference that will allow the proof of theorems about the functions defined in the logic.

2.1. Syntax, Primitive Data Types and Conventions

Because the pure Lisp syntax is unfamiliar to many, we here adopt a more conventional syntax supported by Nqthm-1992's "infix prettyprinter." We will explain the syntax as we go. Case is ignored in the Nqthm syntax. Thus, 'FACT', 'Fact' and 'fact' are the same function symbol. When we talk about a function symbol in running text, as to say "The function 'fact' takes one argument" we generally enclose it in single quotation marks. An exception to this rule is that we sometimes refer to 'fm9001' simply as FM9001. Generally, function symbols, such as 'fact', are set in Roman font, variable symbols, such as x and max, are set in italics, and constants are either set in bold face, such as t, or in Courier, such as `'(t f 123)`, depending on the context. All explicit constants are preceded by a single quotation mark, as above, with a few exceptions noted below. Most function applications are written in the traditional notation, with arguments enclosed in parentheses and separated by commas. Some function symbols, noted explicitly below, are written in an infix syntax. When a constant function, that is, a function taking no arguments, is applied, we do not write the empty pair of parentheses.

The Nqthm logic supports several primitive data types and there is syntactic support for each of them.

Booleans. There are two distinct Boolean constants, called "true" and "false" and written t and f. When we treat an arbitrary x as though it were a Boolean we mean the proposition "$x \neq f$." The most common use of this convention is to say something like "x is true" to mean "$x \neq f$."

Naturals Numbers. The nonnegative integers or "natural numbers" are written in standard decimal notation. Examples include 0, 15, and 435. It is not necessary to quote them. That is, `'15` is the same as 15. We sometimes treat an arbitrary x as though it were a natural number. In such cases, if x is not a natural number, 0 should be used in its place. Most of Nqthm's arithmetic primitives treat their arguments as natural numbers. For example, if x is t, then $x + y$ is the same thing as $0 + y$.

Negative Integers. The negative integers are written in signed decimal notation. It is not necessary to quote them. Thus, `-3` and `-435` are negative integers.

Literal Atoms. The "literal atoms" or "symbols" of Nqthm are constants representing words. Examples include `'nat`, `'halt`, and `'add-nat`. The single quotation mark preceding such a constant is necessary so that the constant is not confused with a variable symbol (when fonts are ignored). Thus, `'nat` is a literal atom constant while *nat* is a variable symbol. An exception to this rule is the constant `'nil` which may be written without the quotation mark, e.g., **nil**. Case is unimportant. Thus, `'NAT` and `'nat` are two ways of writing the same symbol.

Ordered Pairs. Ordered pairs are written as in pure Lisp. For example, the pair consisting of `1` and `0`, which in a conventional mathematics textbook would be written as <1, 0>, is here written as `'(1 . 0)`. The ordered pair <3, <2, <1, 0>>> is written `'(3 2 1 . 0)`. This syntax supports the convention of using ordered pairs to represent lists. The literal atom **nil** is often used to represent the empty list. A nest of ordered pairs may be regarded as a binary tree. When the rightmost leaf of that tree is **nil**, that **nil** is generally not printed in the parenthesized display of the structure. Thus, <1, **nil**> may be written as `'(1 . nil)` but is more often written as `'(1)`. Similarly, <3, <2, <1, **nil**>>> may be written as `'(3 2 1 . nil)` but is more often written as `'(3 2 1)`.

We sometimes treat an arbitrary x as though it were a list. If the value of x is the ordered pair <u, v>, then when we treat x as a list, we treat it as the list whose first element is u and whose remaining elements are in the "list" v. If the value of x is not an ordered pair, then when x is treated as a list it is treated as though it were the empty list.

2.2. Primitive Function Symbols

The axioms of Nqthm-1992 define 62 primitive function symbols, only half of which are used in this work. Here is a brief description of each of the relevant function symbols, divided more or less arbitrarily into groups. We also note below the abbreviation conventions provided by Nqthm. Readers familiar with logic should understand that, except for the abbreviations noted below, all the symbols introduced below are axiomatized in Nqthm as *function symbols* (not *operators* or *relations*).

2.2.1. If and Case

The expression "**if** x **then** y **else** z **endif**" is the most primitive logical expression. Its value is y if x is true and is z otherwise. Note that x is here treated as a proposition and thus by "x is true" we mean "$x \neq$ **f**." Nested **if**-expressions are so common we abbreviate them in the obvious way, as in "**if** p **then** x **elseif** q **then** y **else** z **endif**." Finally,

case on a:
case = key_1 **then** $term_1$

...
case $= key_n$ **then** $term_n$
otherwise $term_{n+1}$ **endcase**

is an abbreviation for

if $a = \, ' key_1$ **then** $term_1$

...
elseif $a = \, ' key_n$ **then** $term_n$
else $term_{n+1}$ **endif**.

2.2.2. *Other Logical Functions*

truep (x)	if x is **t**, then **t**; otherwise **f**.
falsep (x)	if x is **f**, then **t**; otherwise **f**.
$x = y$	if x and y are the same object, then **t**, otherwise **f**.
$p \wedge q$	if p and q are both true, then **t**, otherwise **f**.
$p \vee q$	if p or q is true, then **t**, otherwise **f**.
$\neg p$	if p is true, then **f**, otherwise **t**.
$p \rightarrow q$	if p and q are true or if p is false, then **t**, otherwise **f**.

2.2.3. *Natural Arithmetic*

$x \in N$	if x is a natural number, then **t**, otherwise **f**. Despite the use of the set membership symbol, "$\in$" and the apparent reference to the infinite set **N** of naturals, the Nqthm logic does not include set theory and "$\in N$" is here used as an atomic symbol. Actually, "$x \in N$" might be more clearly understood as "natural-numberp (x)".
fix (x)	if x is a natural number, then x, otherwise **0**. Using our convention for treating arbitrary terms as natural numbers, another way to say this is that fix returns the natural number x. Thus, fix (**7**) is **7**, fix (**-23**) is **0**, and fix (**'abc**) is **0**.
$1 + x$	the natural number one greater than the natural number x. Thus $1 + 3$ is **4**. Since **'abc** is not a natural number, $1 +$ **'abc** is $1 + 0$ which is **1**. Similarly, but perhaps even more surprising, $1 + -3$ is **1**.
$x - 1$	if the natural number x is **0**, then **0**, otherwise one less than the natural number x.
$x \cong 0$	if the natural number x is **0**, then **t**, otherwise **f**.
$x < y$	if the natural number x is less than the natural number y, then **t**, otherwise **f**.
$x + y$	the sum of the natural numbers x and y.

$x - y$	the difference of the natural numbers x and y, unless that difference is negative, in which case the result is **0**.
$x \times y$	the product of the natural numbers x and y.
x **mod** y	the remainder of the natural number x divided by the natural number y. Thus **26 mod 8** is **2**.
x / y	the floor of the quotient of the natural number x divided by the natural number y. Thus **26 / 8** is **3**.
negative-guts (x)	if x is a negative integer, then the absolute value of x, otherwise **0**.
negativep (x)	if x is a negative integer, then **t**, otherwise **f**.
$- x$	the negative of the natural number x. Thus $-$ **23** is **-23**.

2.2.4. List Processing

listp (x)	if x is an ordered pair, then **t**, otherwise **f**.
cons (x, y)	the ordered pair $<x, y>$. Thus, cons $(3,$ **' (2 1)**$)$ is **' (3 2 1)**.
car (x)	if x is the ordered pair $<u, v>$, then u, otherwise **0**. Thus, car $($**' (3 2 1)**$)$ is **3**.
cdr (x)	if x is the ordered pair $<u, v>$, then v, otherwise **0**. Thus, cdr $($**' (3 2 1)**$)$ is **' (2 1)**.
cadr (x), cddr (x), caddr (x), etc	
	When a symbol beginning with the letter 'c,' ending with the letter 'r' and otherwise containing only the letters 'a' and 'd' is used as a function symbol, it is an abbreviation for the nest of 'car's and 'cdr's indicated by the interior letters. Thus, cadr (x) is an abbreviation for car $($cdr $(x))$ and caddadr (x) is an abbreviation for car $($cdr $($cdr $($car $($cdr $(x))))))$. For example, caddr $($**' (0 1 2 3 4)**$)$ is **2**.
list $(x_1, x_2, ..., x_n)$	an abbreviation for cons $(x_1,$ cons $(x_2, ...$ cons $(x_n,$ **nil**$)...))$.
list* $(x_1, x_2, ..., x_n)$	an abbreviation for cons $(x_1,$ cons $(x_2, ...$ cons $(x_{n-1}, x_n)...))$.
nlistp (x)	if x is not an ordered pair, the result is **t** and is **f** otherwise. Thus, nlistp $($**nil**$)$ is **t** and so is nlistp $($**0**$)$.
append (x, y)	the concatenation of the list x with the list y. Thus, append $($**' (5 4)**$,$ **' (3 2 1)**$)$ is **' (5 4 3 2 1)**.
$x \in y$	if x is an element of the list y, then **t**, otherwise **f**.
strip-cars (x)	the list obtained by applying car to each element of the list x and collecting the results. Thus, strip-cars $($**' ((a . 1) (b . 2) (c . 3))**$)$ is **' (a b c)**.
assoc (x, y)	the first element in the list y whose car is x. Thus, assoc $($**'b, ' ((a . 1) (b . 2) (c . 3) (b . 4))**$)$ is **' (b . 2)**.

2.2.5. *Literal Atoms*

litatom (x) if x is a literal atom, then **t**, otherwise **f**.

pack (x) the literal atom "corresponding" to x. The correspondence, not described here, is based on the ASCII assignment of natural numbers to upper case alphabetic characters and certain signs and digits. For example, since the ascii codes for **A**, **B** and **C** are **65**, **66**, and **67**, respectively, pack(**' (65 66 67 . 0)**) is **'ABC**, which may also be written **'abc**.

unpack (x) if x is a literal atom, then the result is an object u such that pack (u) is x. Otherwise the result is **0**. Thus unpack(**'abc**) is **' (65 66 67 . 0)**.

2.3. Let Notation

The notation

let v_1 **be** t_1,
...
      v_n **be** t_n **in**
body **endlet**

is an abbreviation for the term obtained by replacing in *body* the v_i by the corresponding t_i. Thus,

let x **be** $j \times k$,
      y **be** strip-cars (a) **in**
cons $(x,$ cons $(x, y))$ **endlet**

is an abbreviation for cons $(j \times k,$ cons $(j \times k,$ strip-cars $(a)))$.

2.4. Recursive Definitions

The Nqthm logic permits the addition of new axioms defining functions. Certain restrictions, not discussed here, are imposed to insure that inconsistencies are not introduced into the logic. All of the definitions in this book have been proved to meet the restrictions and are admissible. We exhibit a few definitions here simply to introduce the syntax.

Here is a definition of the factorial function.

DEFINITION:
fact $(n) = $ **if** $n \cong 0$ **then 1 else** $n \times$ fact $(n-1)$ **endif**

Thus, for example, fact $(4) = $ **24**.

Technically, the definition of 'fact' is an axiom and fact $(4) = $ **24** is a theorem that can be proved by appealing to rules of inference (such as that every instance of an axiom is a theorem and thus if we replace n in the axiom above by **4** a theorem results), and axioms (such as that $(1+x) \neq$ **0**). We do not discuss such low level proofs here.

Another simple definition is that of the function 'length'.

DEFINITION:
length $(x) = $ **if** nlistp (x) **then** 0 **else** 1+ length (cdr (x)) **endif**

This may be paraphrased as saying "the length of an empty list is 0 and the length of a non-empty list is one greater than the length of its cdr." Thus, length $('$ (**a b c**)) = 3.

2.5. User-Defined Data Types

Nqthm provides a principle, called the "shell principle," with which the user may extend the theory by the addition of axioms defining new inductively constructed data types. Slight use is made of the shell principle in the Piton work and we therefore only describe a limited form of it.

When we say

Add the shell '*const*',
with recognizer function symbol '*recog*',
and n accessors 'ac_1', ..., 'ac_n'.

it means that we are extending the theory to include a new data type. The new type of objects are constructed by the function symbol '*const*', which takes n arguments. The sense in which this type is "new" is that objects constructed by '*const*' are not Booleans, numbers, literal atoms, conses, or any previously mentioned shell. Objects of this new type are recognized by the unary function symbol '*recog*', which returns **t** or **f** according to whether its argument is of the new type, i.e., was constructed by '*const*'. An object of this new type may be thought of as an n-tuple containing the n arguments passed to '*const*' to construct the object. The n accessor function symbols may be used to recover these n components from such an object. That is, for each i between 1 and n we have an axiom of the form

AXIOM:
$ac_i (const(x_1, ..., x_n)) = x_i$

Finally, if an 'ac_i' is applied to something other than an object of this new type, the result is (arbitrarily) 0.

The astute reader might notice that, except for the requirements of "newness," some of our primitive data type functions could have been axiomatized by the use of the shell principle. For example, 'cons' is a shell constructor, with recognizer 'listp' and two accessors, 'car' and 'cdr'.

One example of the use of shells is to represent the state of the formal Piton machine. The incantation

SHELL DEFINITION:

Add the shell 'p-state' of 9 arguments, with
recognizer function symbol 'p-statep', and
accessors 'p-pc',
        'p-ctrl-stk', 'p-temp-stk',
        'p-prog-segment', 'p-data-segment',
        'p-max-ctrl-stk-size', 'p-max-temp-stk-size',
        'p-word-size', and 'p-psw'.

adds to the logic the axioms defining 'p-state' as a function of 9 arguments which constructs 9-tuples of a "new" type. Thus, p-state $(pc, cl, tp, pg, dt, mxc, mxt, w, psw)$ is an object of type 'p-statep' and hence is a 9-tuple. The components can be accessed via the corresponding accessors. Thus, if x is the p-state above, p-pc (x) is pc and p-temp-stk (x) is tp.

3

An Informal Sketch of Piton

Piton is a high-level assembly language for a stack machine. Among the features provided by Piton are:

- execute-only program space
- named read/write global data spaces randomly accessed as one-dimensional arrays
- recursive subroutine call and return
- provision of named formal parameters and stack-based parameter passing
- provision of named temporary variables allocated and initialized to constants on call
- a user-visible temporary stack
- seven abstract data types:
    - integers
    - natural numbers
    - bit vectors
    - Booleans
    - data addresses
    - program addresses (labels)
    - subroutine names
- stack-based instructions for manipulating the various abstract objects
- standard flow-of-control instructions
- instructions for determining resource limitations

As will become apparent when we describe the host machine, FM9001, Piton should not be thought of as an assembly language for FM9001. It is considerably higher level than that.

3.1. An Example Piton Program

We begin our presentation of Piton with a simple example. Below we exhibit a Piton program named **demo**. The program is a list constant in the Nqthm logic [5] and is displayed in the traditional Lisp-like notation. Comments are written in the right-hand column, bracketed by the comment delimiters semi-colon and end-of-line. The **demo** program has three formal parameters, **x**, **y**, and **z**, and two temporary variables, **a** and **i**. The body of **demo** consists of four Piton instructions.

```
(demo (x y z)                         ; formals x, y, and z
      ((a (int -1))                   ; temporary a, initial value -1
       (i (nat 2)))                   ; temporary i, initial value 2
      (push-local y)                  ; push the value of y
      (push-constant (nat 4));        push the natural number 4
      (add-nat)                       ; add the top two items
      (ret))                          ; return
```

Piton has a user-visible stack that is used to pass actuals to primitive operators as well as to user-defined subroutines such as **demo**. When **demo** is called, as by executing the Piton instruction **(call demo)**, the topmost three items from Piton's stack are popped off and used as the actual values of the formals **x**, **y**, and **z**. In addition, the temporary variable **a** is initialized to the integer -1 and **i** is initialized to the natural number 2. The values of all five of these "local" variables are restored when **demo** returns to its caller.

The body of **demo** has four Piton instructions in it. The first, **(push-local y)**, pushes the value of the local variable **y** onto the temporary stack. The second, **(push-constant (nat 4))**, pushes the natural number 4 onto the temporary stack. The third, **(add-nat)**, pops the topmost two items off the temporary stack, adds them together (expecting both to be naturals), and pushes the result onto the temporary stack. The last instruction returns control to the calling environment. The sum just computed is on top of the stack and is considered the result. In summary, this silly program adds 4 to the value of its second argument and ignores the other arguments. Its two temporary variables are not used.

Now consider the following sequence of Piton instructions.

```
(push-constant (addr (delta1 . 25)))
(push-constant (nat 17))
(push-constant (bool t))
(call demo)
```

This sequence pushes three items onto the stack and then calls **demo**. The **call** pops the three objects off the stack and uses them as the actuals. **Demo**'s first formal, **x**, is bound to the data address **(delta1 . 25)** — the address of the 25th location of the global array named **delta1**. **Demo**'s second argument, **y**, is bound to the natural number 17. Its third argument, **z**, is bound to the Boolean value **t**. The execution of **demo** pushes 21 (the sum of 17 and 4) and returns. Thus, the net effect of the four instructions above — barring a variety of runtime errors such as stack overflow — is to push a 21 onto the stack.

3.2. Piton States

The Piton machine is a fairly conventional stack based von Neumann state transition machine with an execute-only program memory. Roughly speaking, a particular instruction is singled out as the "current instruction" in any Piton state. When "executed" each instruction changes the state in some way, including changing the identity of the current instruction. The Piton machine operates on an initial state by iteratively executing the current instruction until some termination condition is met.

A Piton state, or *p-state*, is a 9-tuple. Formally, a p-state is a new user-defined data type introduced into Nqthm with the shell principle. P-states are constructed by the 9-argument function 'p-state'; each component of the resulting 9-tuple is accessed by a function naming the component. We give the function names below as we enumerate the components of a p-state.

- a *program counter* (accessed via the function 'p-pc'), indicating which instruction in which subroutine is the next to be executed;

- a *control stack* ('p-ctrl-stk'), recording the hierarchy of subroutine invocations leading to the current state;

- a *temporary stack* ('p-temp-stk'), containing intermediate results as well as the arguments and results of subroutine calls;

- a *program segment* ('p-prog-segment'), defining a system of Piton programs or subroutines;

- a *data segment* ('p-data-segment'), defining a collection of disjoint named indexed data spaces (i.e., global arrays);

- a *maximum control stack size* ('p-max-ctrl-stk-size');

- a *maximum temporary stack size* ('p-max-temp-stk-size');

- a *word size* ('p-word-size'), governing the size of numeric constants and bit vectors; and

- a *program status word* ('p-psw') usually just called the *psw*.

We put a variety of additional restrictions on the components of a p-state. For example, we require that every instruction in every program is syntactically well-formed and mentions no variables other than the locals of the containing program or the globals declared in the data segment. We also require that every data object occurring in the state is compatible with the state, e.g., every object tagged "address" is a legal address in that state, etc. We call such p-states *proper p-states*. The formalization of this syntactic concept is embodied in the function 'proper-p-statep' which is defined on page 237.

The program counter of a p-state names one of the programs in the program segment, which we call the *current program*, and gives the position of one of the instructions in that program's body, which we call the *current instruction*. We say *control* is *in* the current program and *at* the current instruction.

The control stack of the p-state is a stack of *frames*, the topmost frame describing the currently active subroutine invocation and the successive frames describing the hierarchy of suspended invocations. The topmost frame is the only frame directly accessible to Piton instructions. Each frame has two fields in it. One contains the *bindings* of the local variables of the invoked program. The other contains the *return program counter*, which is the program counter to which control is to return when the subroutine exits.

Recall from the discussion of **demo** that Piton subroutines have formal parameters, temporary variables, and then a body consisting of optionally labeled Piton instructions. When a subroutine is *called* or *invoked* the actual parameters of the subroutine are passed via the temporary stack. Upon call, a new frame is pushed onto the control stack. The actuals are removed from the temporary stack and the formals are bound to those actuals in the new frame. The temporary variables are also bound in the new frame. Then control is transferred to the first instruction in the body of the subroutine. All references to local variables in the instructions of the called subroutine refer implicitly to the current bindings. When the return instruction is executed, the subroutine returns to its caller. The top frame of the control stack is popped off, thus restoring the caller's locals. In short, the values assigned to the local variables of a subroutine are local to a particular invocation and cannot be accessed or changed by any other subroutine or recursive invocation. We define "local variables" and what we mean by the "appropriate values" when we discuss Piton programs.

3.3. Type Checking

Piton programs manipulate seven types of data: integers, natural numbers, Booleans, fixed length bit vectors, data addresses, program addresses, and subroutine names.

All objects are "first class" in the sense that they can be passed around and stored into arbitrary variable, stack, and data locations. *There is no type checking in the Piton syntax.* A variable can hold an integer value now and a Boolean value later, for example.

Each type comes with a set of Piton instructions designed to manipulate objects of that type. For example, the **add-nat** instruction adds two naturals together to produce a natural; the **add-addr** instruction increments a data address by a natural to produce a new data address. If the "dynamic restrictions" on an instruction are violated at runtime, e.g., if **add-nat** is executed on a natural and a Boolean, the semantics of Piton defines the resulting state to be "erroneous" and so marks the state by an appropriate setting of the psw. Arrival at an erroneous state effectively halts the Piton machine.

However, our compiler for Piton does not include any treatment of error checking. The compiler is limited in the sense that it can only correctly compile non-erroneous programs.

Such cavalier runtime treatment of types—i.e., no syntactic type checking and no runtime type checking—would normally be an invitation to disaster. In most programming languages the definition of the language is embedded in only two mechanical devices: the compiler (where syntactic checks are made) and the runtime system (where semantic checks are made). If some feature of the language (e.g., correct use of the type system) is not checked by either of these two devices, the programmer bears a heavy responsibility and must be very careful.

But the Piton programmer is relieved of this burden by an unconventional third mechanical device: the mechanized formal semantics. This device—actually the Nqthm theorem prover initialized with the formal definition of Piton—completely embodies the formal semantics of Piton. If a programmer wishes to establish that a program is non-erroneous a mechanically checked proof of that assertion can be undertaken.

As programmers we find this a refreshing state of affairs. We are relieved of the burden of syntactic restrictions in the language—objects can be slung around any way we please. We are relieved of the inefficiency of checking types at runtime. But we don't have to worry about having made mistakes. The price, of course, is that we must be willing to prove our programs correct.

3.4. Data Types

As noted, Piton supports seven primitive data types. The syntax of Piton requires that all data objects be tagged by their type. Thus, (**int** 5) is the way we write the integer 5, while (**nat** 5) is the way we write the natural number 5. The question ''are they the same?'' cannot arise in Piton because no operation compares them.

Below we characterize all of the legal instances of each type. However, this must be done with respect to a given p-state, since the p-state determines the resource limitations, legal addresses, etc. Let w be the word size of the p-state implicit in our discussion. In the examples of this section we assume w is **8**. Our FM9001 implementation of Piton is for word size **32**. The formalization of the concept of ''legal Piton data object'' is embodied in the function 'p-objectp' which is defined on page 208.

3.4.1. Integers

Piton provides the integers, i, in the range $-2^{w-1} \leq i < 2^{w-1}$. We say such integers are *representable* in the given p-state. Observe that there is one more representable negative integer than representable positive integers. Integers are written down in the form (**int** i), where i is an optionally signed integer in decimal notation. For example, (**int** **-4**) and (**int** **3**) are Piton integers. Piton provides instructions for adding, subtracting, and comparing integers. It is also possible to convert non-negative integers into naturals.

3.4.2. Natural Numbers

Piton provides the natural numbers, n, in the range $0 \leq n < 2^w$. We say such naturals are *representable* in the given p-state. Naturals are written down in the form (**nat** *n*), where *n* is an unsigned integer in decimal notation. For example, (**nat 0**) and (**nat 7**) are Piton naturals. Piton provides instructions for adding, subtracting, doubling, halving, and comparing naturals. Naturals also play a role in those instructions that do address manipulation, random access into the temporary stack, and some control functions.

3.4.3. Booleans

There are two Boolean objects, called **t** and **f**. They are written down (**bool t**) and (**bool f**).[3] Piton provides the logical operations of conjunction, disjunction, negation and equivalence. Several Piton instructions generate Boolean objects (e.g., the "less than" operators for integers and naturals).

3.4.4. Bit Vectors

A Piton bit vector is an array of **1**'s and **0**'s as long as the word size, *w*, of the Piton state. Bit vectors are written in the form (**bitv** *v*) where *v* is a list of length *w*, enclosed in parentheses, containing only 1's and 0's. For example (**bitv (1 1 1 1 0 0 0 0)**) is a bit vector when *w* is **8**. Operations on bit vectors include componentwise conjunction, disjunction, negation, exclusive-or, left and right shift, and equivalence.

3.4.5. Data Addresses

A Piton data address is a pair consisting of a name and a natural number. To be legal in a given p-state, the name must be the name of some data area in the data segment of the state and the number must be less than the length of the array associated with the named data area. Data addresses are written (**addr** (*name* . *n*)). Such an address refers to the nth element of the array associated with *name*, where enumeration is 0 based, starting at the left hand end of the array. For example, if the data segment of the state contains a data area named **delta1** that has an associated array of length 128, then (**addr** (**delta1** . **122**)) is a data address. The operations on data addresses include incrementing, decrementing, and comparing addresses, fetching the object at an address, and depositing an object at an address.

---

[3]Note to those familiar with the Nqthm logic: The **t** and **f** used in the representation of the Piton Booleans are *not* the t and f of the logic but the literal atoms **'t** and **'f** of the logic.

3.4.6. Program Addresses

A Piton program address is a pair consisting of a name and a natural number. To be legal in a given p-state, the name must be the name of some program in the program segment of the state and the number must be less than the length of the body of the named program. Program addresses are written **(pc** (*name* **.** *n***))**. Such an address refers to the nth instruction in the body of the program named *name*, where enumeration is 0 based starting with the first instruction in the body. For example, if the program segment of the state contains a program named **setup** that has 200 instructions in its body, then **(pc (setup . 27))** is a legal program address. Program addresses can be compared and control can be transferred to (the instruction at) a program address. Some instructions generate program addresses. But it is impossible to deposit anything at a program address (just as it is impossible to transfer control to a data address).

The program counter component of a p-state is an object of this type. For example, to start a computation at the first instruction of the program named **main**, the program counter in the state should be set to **(pc (main . 0))**.

3.4.7. Subroutines

A Piton subroutine name is just a name. To be legal, it must be the name of some program in the program segment. Subroutine names are written **(subr** *name***)**. For example, if **setup** is the name of a program in the program segment, then **(subr setup)** is a subroutine object in Piton. The only operation on subroutine objects is to call them.

3.5. The Data Segment

The Piton data segment contains all of the global data in a p-state. The data segment is a list of *data areas*. Each data area consists of a literal atom *data area name* followed by one or more Piton objects, called the *array* associated with the name. The objects in the array are implicitly indexed from 0, starting with the leftmost. Using data addresses, which specify a name and an index, Piton programs can access and change the elements in an array.

We sometimes call a data area name a *global variable*. Some Piton instructions expect global variables as their arguments and operate on the 0th position of the named data area. We define the *value* of a global variable to be the contents of the 0th location in its associated array. This is a pleasant convention if the data area only has one element but tends to be confusing otherwise.

Here, for example, is a data segment:

```
((len (nat 5))
 (a   (nat 0)
      (nat 1)
      (nat 2)
      (nat 3)
      (nat 4))
 (x   (int -23)
      (nat 256)
      (bool t)
      (bitv (1 0 1 0 1 1 0 0))
      (addr (a . 3))
      (pc (setup . 25))
      (subr main))).
```

This segment contains three data areas, **len**, **a**, and **x**. The **len** area has only one element and so is naturally thought of as a global variable. Its value is the natural number 5. The **a** array is of length 5 and contains the consecutive naturals starting from 0. While **a** is of homogeneous type as shown, Piton programs may write arbitrary objects into **a**. The third data area, **x**, has an associated array of length 7. It happens that this array contains one object of every Piton type.

Let **addr** be the Piton data address object **(addr (x . 1))**. If we fetch from **addr** we get **(nat 256)**. If we deposit **(nat 7)** at **addr** the data segment becomes

```
((len (nat 5))
 (a   (nat 0)
      (nat 1)
      (nat 2)
      (nat 3)
      (nat 4))
 (x   (int -23)
      (nat 7)
      (bool t)
      (bitv (1 0 1 0 1 1 0 0))
      (addr (a . 3))
      (pc (setup . 25))
      (subr main))).
```

If we increment **addr** by one and then fetch from **addr** we get **(bool t)**.

The individual data areas are totally isolated from each other. Despite the fact that addresses can be incremented and decremented, there is no way for a Piton program to manipulate **addr**, which addresses the area named **x**, so as to obtain an address into the area named **a**.

3.6. The Program Segment

The program segment of a p-state is a list of "program definitions". A *program definition* (or, interchangeably, a *subroutine definition*) is an object of the following form

$$(name \ (v_0 \ v_1 \ ... \ v_{n-1})$$
$$((v_n \ i_n) \ (v_{n+1} \ i_{n+1}) \ ... \ (v_{n+k-1} \ i_{n+k-1}))$$
$$ins_0$$
$$ins_1$$
$$...$$
$$ins_m),$$

where *name* is the *name* of the program and is some literal atom; $v_0, ..., v_{n-1}$ are the $n \geq 0$ *formal parameters* of the program and are literal atoms; $v_n, ..., v_{n+k-1}$ are the $k \geq 0$ *temporary variables* of the program and are literal atoms; $i_n, ..., i_{n+k-1}$ are the *initial values* of the corresponding temporary variables and are data objects in the state in which the program occurs; and $ins_0, ... ins_m$ are $m+1$ optionally labeled Piton instructions, called the *body* of the program. The body must be non-empty.

The *local variables* of a program are the formal parameters together with the temporary variables. The values of the local variables of a subroutine may be accessed and changed by position as well as by name. For this purpose we enumerate the local variables starting from 0 in the same order they are displayed above.

As noted previously, upon subroutine call a stack frame is created for that invocation and the local variables of the called subroutine are bound in that frame. The n formal parameters are initialized from the temporary stack. The topmost n elements of the temporary stack are called the *actuals* for the call. They are removed from the temporary stack and associated with the formal parameters in reverse order. That is, the first actual pushed onto the temporary stack (i.e., the deepest) is the value associated with the first formal parameter. The k temporary variables are bound to their respective initial values at the time of call.

The instructions in the body of a Piton program may be optionally labeled. A *label* is a literal atom. To attach label *lab* to an instruction, *ins*, we write

(dl *lab comment ins*)

where *comment* is any object in the logic and is totally ignored by the Piton semantics and implementation. We say *lab* is *defined* in a program if the body of the program contains (DL *lab* ...) as one of its members. Such a form is called a *def-label form* because it defines a label. Label definitions are local to the program in which they occur. Use of the atom loop, for example, as a label in some instruction in a program refers to the (first) point in that program at which loop is defined in a def-label form.

Because of the local nature of label definitions it is not possible for one program to jump to a label in another. A similar effect can be obtained efficiently using the data objects of type pc. The popj instruction transfers control to the program address on the top of the temporary stack—provided that address is in the current program.[4] Thus, if subroutine **master** wants to jump to the 22nd instruction of subroutine **slave**, it could **call slave** and pass it the argument **(pc (slave . 22))**, and **slave** could do a **popj** as its first instruction to branch to the desired location.

The last instruction, ins_m must be a return or some form of unconditional jump. It is not permitted to "fall off" the end of a Piton program.

The formalization of the concept of a syntactically well-formed Piton program is embodied in the function 'proper-p-programp' which is defined on page 235. Part of the constraints on proper p-states is that they contain syntactically well-formed Piton programs.

3.7. Instructions

Figure 3-1 lists the Piton instructions. The instructions are organized informally into groups (and some instructions are included in more than one group). The language as currently defined provides 65 instructions. We do not regard the current instruction set as fixed in granite; we imagine Piton will continue to evolve to suit the needs of its users.

We describe selected instructions informally below. For a summary of each Piton instruction, see Appendix I, where we also explain in more detail the conventions used below. "*Syn*" stands for "Syntactic Restrictions" on the instruction form, "*Pre*" stands for "Dynamic Precondition" and the page numbers indicate where the informal remarks are expressed formally. Unless otherwise indicated, every instruction increments the program counter by one so that the next instruction to be executed is the instruction following the current one in the current subroutine. All references to "the stack" refer to the temporary stack unless otherwise specified. When we say "push" or "pop" without mentioning a particular stack we mean to push or pop the temporary stack.

The style of our semantics can be appreciated by following the page references in these summaries. The **jump-case** instruction is typical. Here is the informal summary.

(**jump-case** lab_0 lab_1 ... lab_n)
> *Syn* (page 233): Each of the lab_i is a label in the containing program. *Pre* (page 201): There is a natural, i, on top of the stack and $i \leq n$. *Effects* (page 201): Pop once and then jump to lab_i.

The page numbers make it easy to relate this informal description to the formal semantics of the instruction. As it says above, the syntactic restriction is defined formally on page 233 where we find the definition of the following predicate.

---

[4]There is no way, in Piton, to transfer control into another subroutine except via the call/return mechanism.

Control	Integers	Natural Numbers
call	add-int	add-nat
jump	add-int-with-carry	add-nat-with-carry
jump-case	add1-int	add1-nat
no-op	eq	div2-nat
ret	int-to-nat	eq
test-bitv-and-jump	lt-int	lt-nat
test-bool-and-jump	neg-int	mult2-nat
test-int-and-jump	sub-int	mult2-nat-with-carry-out
test-nat-and-jump	sub-int-with-carry	sub-nat
	sub1-int	sub-nat-with-carry
		sub1-nat

Variables	Booleans	Bit Vectors
locn	and-bool	and-bitv
pop-global	eq	eq
pop-local	not-bool	lsh-bitv
pop-locn	or-bool	not-bitv
push-global		or-bitv
push-local		rsh-bitv
set-global		xor-bitv
set-local		

Stack	Data Addresses	Subroutines
deposit-temp-stk	add-addr	eq
fetch-temp-stk	deposit	pop-call
pop	eq	
pop*	fetch	
popn	lt-addr	program addresses
push-constant	sub-addr	
push-temp-stk-index		eq
		popj
		pushj

Resources

jump-if-temp-stk-empty
jump-if-temp-stk-full
push-ctrl-stk-free-size
push-temp-stk-free-size

Figure 3-1: Piton Instructions

DEFINITION:
proper-p-jump-case-instructionp (*ins*, *name*, *p*)
=

    listp (cdr (*ins*))
∧ all-find-labelp (cdr (*ins*),
                      program-body (definition (*name*, p-prog-segment (*p*)))))

The predicate checks that *ins* has the syntactic form of a **jump-case** instruction provided the 'car' of *ins* is **'jump-case**. We have not exhibited definitions of all the concepts involved, but this predicate can probably be understood from the informal summary and the names of the functions used. In particular, the predicate checks that the 'cdr' of *ins* is a non-empty list of labels defined in the program in which the instruction occurs. In Appendix II we formally define every function used in the semantics of Piton.

    The dynamic precondition for **jump-case** is defined on page 201.

DEFINITION:
p-jump-case-okp (*ins*, *p*)
=

    listp (p-temp-stk (*p*))
∧ p-objectp-type (**'nat**, top (p-temp-stk (*p*)), *p*)
∧ (untag (top (p-temp-stk (*p*))) < length (cdr (*ins*)))

It checks that the temporary stack in the current p-state is non-empty, that the top of the stack is a Piton object of type **nat** that is less than the number of labels in the **jump-case** instruction. 'Untag', defined on page 241, strips off the tag of a Piton object.

    Finally, the effects of the **jump-case** instruction are specified on page 201.

DEFINITION:
p-jump-case-step (*ins*, *p*)
=

p-state (pc (get (untag (top (p-temp-stk (*p*))), cdr (*ins*)), p-current-program (*p*)),
        p-ctrl-stk (*p*),
        pop (p-temp-stk (*p*)),
        p-prog-segment (*p*),
        p-data-segment (*p*),
        p-max-ctrl-stk-size (*p*),
        p-max-temp-stk-size (*p*),
        p-word-size (*p*),
       'run)

This function defines the p-state that results from executing a **jump-case** instruction on state *p*. Observe that the program counter of the new p-state is set to the **pc** corresponding to the appropriate label and the temporary stack is popped.

    We now exhibit informal summaries of a few other Piton instructions to indicate some features of the language. A complete collection of summaries may be found in Appendix I.

  (add-addr)   *Syn* (page 227): None. *Pre* (page 185): There is a natural, *n*, on top of the stack and a data address, *a*, immediately below it. The result of incrementing *a* by *n* is a legal data address. *Effects* (page 185): Pop twice and then push the data address obtained by incrementing a by n.

(add-int)   *Syn* (page 228): None. *Pre* (page 186): There is an integer, i, on top of the stack and an integer, j, immediately below it. The integer sum of i and j is representable. *Effects* (page 186): Pop twice and then push the integer sum of i and j.

(add-int-with-carry)

Syn (page 228): None. *Pre* (page 186): There is an integer, i, on top of the stack, an integer, j, immediately below it, and a Boolean, c, below that. *Effects* (page 187): Pop three times. Let k be **1** if c is **t** and **0** otherwise. Let *sum* be the integer sum $i + j + k$. If *sum* is representable in the word size, w, of this p-state, push the Boolean **f** and then the integer *sum*; if *sum* is not representable and is negative, push the Boolean **t** and the integer $sum + 2^w$; if sum is not representable and positive, push the Boolean **t** and the integer $sum - 2^w$.

(call *subr***)**   *Syn* (page 228): *Subr* is the name of a program in the program segment. *Pre* (page 190): Suppose that *subr* has n formal variables and k temporary variables. Then the temporary stack must contain at least n items and the control stack must have at least **2** $+ n + k$ free slots. *Effects* (page 191): Transfer control to the first instruction in the body of *subr* after removing the topmost n elements from the temporary stack and constructing a new frame on the control stack. In the new frame the formals of *subr* are bound to the n elements removed from the temporary stack, in reverse order, the temporaries of *subr* are bound to their declared initial values, and the return program counter points to the instruction after the **call**.

(deposit)   *Syn* (page 229): None. *Pre* (page 192): There is a data address, a, on top of the stack and an arbitrary object, *val*, immediately below it. *Effects* (page 192): Pop twice and then deposit *val* into the location addressed by a.

(deposit-temp-stk)

Syn (page 229): None. *Pre* (page 192): There is a natural number, n, on top of the stack and some object, *val*, immediately below it. Furthermore, n is less than the length of the stack after popping two elements. *Effects* (page 192): Pop twice and then deposit *val* at the nth position in the temporary stack, where positions are enumerated from **0** starting at the bottom.

(eq)   *Syn* (page 229): None. *Pre* (page 193): The temporary stack contains at least two items and the top two are of the same type. *Effects* (page 193): Pop twice and then push the Boolean **t** if they are the same and the Boolean **f** if they are not.

(fetch)   *Syn* (page 229): None. *Pre* (page 193): There is a data address, *a*, on top of the stack. *Effects* (page 193): Pop once and then push the contents of address *a*.

(int-to-nat) *Syn* (page 232): None. *Pre* (page 200): There is a non-negative integer, *i*, on top of the stack. *Effects* (page 200): Pop and then push the natural *i*.

(jump-if-temp-stk-empty *lab***)**
                    Syn (page 233): *lab* is a label in the containing program. *Pre* (page 201): None. *Effects* (page 201): Jump to *lab* if the temporary stack is empty.

(pop-global *gvar***)**
                    Syn (page 234): *Gvar* is a global variable, i.e., the name of a data area in the data segment of the containing p-state. *Pre* (page 210): There is an object, *val*, on top of the stack. *Effects* (page 210): Pop and assign *val* to the (0th position of the array associated with the) global variable *gvar*.

(pop-local *lvar***)**
                    Syn (page 234): *Lvar* is a local variable of the containing program. *Pre* (page 211): There is an object, *val*, on top of the stack. *Effects* (page 211): Pop and assign *val* to the local variable *lvar*.

(push-constant *const***)**
                    Syn (page 235): *Const* is either a legal Piton object in the containing p-state, the atom **pc**, or a label in the containing program. *Pre* (page 213): There is room to push at least one item. *Effects* (page 213): If *const* is a Piton object, push *const*; if *const* is the atom **pc**, push the program counter of the next instruction; otherwise push the program counter corresponding to the label *const*.

(push-ctrl-stk-free-size)
                    Syn (page 236): None. *Pre* (page 214): There is room to push at least one item. *Effects* (page 214): Push the natural number indicating how many more cells can be created on the control stack before the maximum control stack size is exceeded.

(test-int-and-jump *test lab***)**

Syn (page 238):   *Test* is one of **neg, not-neg, zero, not-zero, pos** or **not-pos**, and *lab* is a label in the containing program. *Pre* (page 225): There is an integer, i, on top of the stack. *Effects* (page 225): Pop once and then jump to *lab* if *test* is satisfied, as indicated below.

test	condition tested
neg	$i < 0$
not-neg	$i \geq 0$
zero	$i = 0$
not-zero	$i \neq 0$
pos	$i > 0$
not-pos	$i \leq 0$

3.8. The Piton Interpreter

Associated with each instruction is a predicate on p-states called the *precondition* for the instruction. This predicate insures that it is legal to execute the instruction in the current p-state, i.e., they are "dynamic" not syntactic or static checks. Generally speaking, the precondition of an instruction checks that the operands exist on the stack, have the appropriate types and do not cause the machine to exceed its resource limits.

Also associated with each instruction is a function from p-states to p-states called the *step* or *effects* function. The step function for an instruction defines the state produced by executing the instruction, provided the precondition is satisfied. Most of the step functions increment the program counter by one and manipulate the stacks and/or global data segment.

The Piton interpreter is a typical von Neumann state transition machine. The interpreter iteratively constructs the new current state by applying the step function for the current instruction to the current state, provided the precondition is satisfied. This process stops, if at all, either when a precondition is unsatisfied or a top-level return instruction is executed.[5] The property of being a proper p-state is preserved by the Piton interpreter. That is, if the initial state is proper, so is the final state. The formalization of the Piton interpreter is the function 'p' which is defined on page 185. The expression p(s, n) is the p-state obtained by executing n instructions starting in p-state s. Generally we use the variable symbol p to denote a p-state and thus the result of stepping p forward n steps is p(p, n). The use of the symbol "p" both as a function symbol and a variable symbol is sometimes confusing but is unambiguous.

We use the function 'p' synonymously with such phrases as "Piton," the "Piton machine" and the "Piton interpreter." By "Piton implementation" we mean the function 'load' which maps a Piton state into an FM9001 state.

---

[5]We formalize this machine constructively by defining the function that iterates the process a given number of times.

3.9. Erroneous States

What does the Piton machine do when the precondition for the current instruction is not satisfied? This brings us to the role of the program status word, psw, in the state and its use in error handling. This has important consequences in the design, implementation, and proof of Piton.

The psw is normally set to the literal atom **run,** which indicates that the computation is proceeding normally. The psw is set to **halt** by the **ret** (return) instruction when executed in the top level program; the **halt** psw indicates successful termination of the computation. The psw is set to one of a many *error conditions* whenever the precondition for the current instruction is not satisfied. Any state with a psw other than **run** or **halt** is called an *erroneous* state. The Piton interpreter is defined as an identity function on erroneous states.

No Piton instruction (i.e., no precondition or step function) inspects the psw. It is impossible for a Piton program to trap or mask an error. The psw and the notion of erroneous states are metatheoretic concepts in Piton; they are used to define the language but are not part of the language.

We consider an implementation of Piton correct if it has the property that it can successfully carry out every computation that produces a non-erroneous state. This is made formal when we present our correctness theorem. But the consequences to the implementation should be clear now. For example, the **add-nat** instruction requires that two natural numbers be on top of the stack and that their sum is representable. This need not be checked at run-time by the implementation of Piton. The run-time code for **add-nat** can simply add together the top two elements of the stack and increment the program counter. If the Piton machine produces a non-erroneous state on the **add-nat** instruction, then the implementation follows it faithfully. If the Piton machine produces an erroneous state, then it does not matter what the implementation does. For example, our implementation of **add-nat** does not check that the stack has two elements, that the top two elements are naturals, or that their sum is representable. It is difficult even to characterize the damage that might be caused if these conditions are not satisfied when our code is executed. As noted in our discussion of type checking, mechanical proof can be used to certify that no such errors occur.

The language contains adequate facilities to program explicit checks for all resource errors. For example, **add-nat-with-carry** will not only add two naturals together, it will push a Boolean which indicates whether the result is the true sum. If you have to test whether the sum is representable, use **add-nat-with-carry.** On the other hand, if you *know* the result is representable, use **add-nat.**

But, unless you are adding constants together, how can you possibly know the result is representable? That is, under what conditions can you to use **add-nat** and still prove the absence of errors? This brings us to the crux of the problem. When you write a Piton program and prove it non-erroneous you do not have to prove the total absence of errors. You *do* have to state the conditions under which the program may be called and prove the absence of errors under those conditions. For example, a typical hypothesis about the initial state might be that the sum of the top two elements of the stack is representable and the stack contains at least five free cells. These conditions are expressed in the logic, not in Piton. We illustrate the handling of errors in the next chapter.

Most of the rest of this book concerns the implementation of Piton on verified hardware and the proof of the correctness of the implementation. With three exceptions, this material is not relevant to the reader who simply wishes to use Piton. The three exceptions are Chapters 4, 6 and II. Chapter 4 illustrates the use of Piton as a verifiable programming language: we specify, implement, and prove the correctness of a big number addition algorithm. Chapter 6 informally describes what was proved about the FM9001 implementation of Piton and can be regarded as the warranty that comes with the FM9001 Piton implementation. Appendix II gives the formal definition of Piton and thus serves as a precise reference manual for the language.

Big Number Addition

In this chapter we consider an example programming problem and its solution in Piton. The problem is to specify, implement, and verify a program for doing "big number addition." This chapter is a rather long but instructive detour from our main goal of implementing Piton on FM9001 and proving the implementation correct.

4.1. An Informal Explanation

A "big number" is a fixed length array of "digits," each digit being a natural number less than a fixed "base." The intended interpretation of such an array is that it represents the natural number obtained by summing the product of the successive digits and successive powers of the base. In our representation of big numbers we put the least significant digit in position 0. For example, a big number array of length 5 representing the number 123 in base ten is '(3 2 1 0 0). Of course, normally the base of a big number system is the first unrepresentable natural on the host machine. For example, in a 32-bit wide machine, the natural base for big number arithmetic is 2^{32}, so that each digit is a full word.

Big number addition is the process that takes as input two big number arrays and produces as output the big number array representing their sum. For example, the table below shows two naturals, their corresponding base one hundred big-number arrays of length 5 and the two sums (the natural sum and the corresponding big number sum).

12, 345, 678	'(78 56 34 12 0)
+ 70, 005, 020	'(20 50 0 70 0)
82, 350, 698	'(98 6 35 82 0)

Given our representation of big numbers one adds corresponding digits starting at the leftmost, carrying to the right.

4.2. A Formal Explanation

A *big number (in base base)* is an object *a* satisfying the predicate bignp (*a*, *base*), where

DEFINITION:
bignp (*a*, *base*)

=

if nlistp (*a*) **then** *a* = **nil**
 else   listp (car (*a*))
       ∧ (type (car (*a*)) = **'nat**)
       ∧ (untag (car (*a*)) ∈ N)
       ∧ (untag (car (*a*)) < *base*)
       ∧ (cddr (car (*a*)) = **nil**)
       ∧ bignp (cdr (*a*), *base*) **endif**

The function 'type' is defined on page 241 as part of the formal definition of Piton. Type (*x*) returns the tag of the Piton object *x*. Thus, the definition of 'bignp' above means that a big number is a proper list of tagged naturals, each of which is less than the base. An example big number in base one hundred is **' ((nat 78) (nat 56) (nat 34) (nat 12) (nat 0))**.

The natural number *represented by* a big number *a* in base *base* is bign⇒nat (*a*, *base*).

DEFINITION:
bign⇒nat (*a*, *base*)

=

if nlistp (*a*) **then** 0
 else untag (car (*a*)) + (*base* × bign⇒nat (cdr (*a*), *base*)) **endif**

For example, the natural represented by **' ((nat 78) (nat 56) (nat 34) (nat 12) (nat 0))** in base one hundred is

$$78 + 100*(56 + 100*(34 + 100*(12 + 100*(0 + 0))))$$

=

$$78 + 56*100 + 34*100^2 + 12*100^3 + 0*100^4$$

=

$$78 + 5600   + 340000   + 12000000$$

=

12345678.

We define big number addition by the pair of functions shown below.

DEFINITION:
big-add-array $(a, b, c, base)$
=
if nlistp (a) **then nil**
 else cons (tag $('$**nat**,
                     untag $(car(a))$ + untag $(car(b))$ + tv$\Rightarrow$nat(c) **mod** $base$),
             big-add-array (cdr (a),
                     cdr (b),
                     (untag $(car(a))$ + untag $(car(b))$ + tv$\Rightarrow$nat(c))
                     $\geq$ $base$,
                     $base$)) **endif**

DEFINITION:
big-add-carry-out $(a, b, c, base)$
=
if nlistp (a) **then** bool (c)
 else big-add-carry-out (cdr (a),
                     cdr (b),
                     (untag $(car(a))$ + untag $(car(b))$ + tv$\Rightarrow$nat(c))
                     $\geq$ $base$,
                     $base$) **endif**

Both functions take as input two big numbers, a and b, an "input carry flag", c, and the specified base, $base$. We assume a and b are both of length n. The first function, 'big-add-array', produces a big number of length n. The second function, 'big-add-carry-out', produces a truth value, called the *carry out*, which indicates whether the sum is too big to be represented in n digits. The functions 'tag' and 'bool' used above are defined on pages 241 and 180 as part of the formal definition of Piton. Tag (n) produces the tagged object $'$ (**nat** n) and bool (c) produces a tagged Piton Boolean from a truth value, e.g., bool (t) is $'$ (**bool** **t**). The subsidiary function 'tv$\Rightarrow$nat' converts a truth value to a natural and is defined as

DEFINITION:
tv$\Rightarrow$nat (c)
=
if c **then 1**
 else 0 endif

Recall the previously shown example of big number addition.

	12, 345, 678	(78	56	34	12	0)
+	70, 005, 020	(20	50	0	70	0)
	82, 350, 698	(98	6	35	82	0)

We display the formal version of this example in two parts. The formula

```
big-add-array (
 '((NAT 78)  (NAT 56)  (NAT 34)  (NAT 12)  (NAT 0)),
 '((NAT 20)  (NAT 50)  (NAT  0)  (NAT 70)  (NAT 0)),
 f, 100)
=
 '((NAT 98)  (NAT  6)  (NAT 35)  (NAT 82)  (NAT 0))
```

exhibits the five digit addition and the formula

```
big-add-carry-out (
 '((NAT 78) (NAT 56) (NAT 34) (NAT 12) (NAT 0)),
 '((NAT 20) (NAT 50) (NAT  0) (NAT 70) (NAT 0)),
 f, 100)
=
 '(bool f)
```

shows that there is no carry out.

We can package 'big-add-array' and 'big-add-carry-out' into a single function, here called 'big-plus', which takes two big numbers of length n and returns the big number sum of length n+1 obtained by concatenating to the value of 'big-add-array' a single high order digit obtained from the carry out.

DEFINITION:
big-plus $(a, b, c, base)$
=
append (big-add-array $(a, b, c, base)$,
         list (tag ('**nat**, bool-to-nat (untag (big-add-carry-out $(a, b, c, base)$))))))

'bool-to-nat' (page 180) is defined as part of the formal definition of Piton and converts the Piton Booleans **t** and **f** to 1 and 0, respectively. On the two input big numbers shown above, 'big-plus' returns '((**nat 98**) (**nat 6**) (**nat 35**) (**nat 82**) (**nat 0**) (**nat 0**)). Note the extra high order 0 indicating that carry out did not occur.

Calling this the ''sum'' of the two big numbers is justified by the observation that the natural represented by the big number sum of a and b is the Peano sum of the naturals represented by a and b. This holds for all big numbers a and b of equal length, provided the base is a natural number greater than 1. The formal rendering of this general remark is

THEOREM:
    (bignp $(a, base)$
  $\wedge$ bignp $(b, base)$
  $\wedge$ (length (a) = length (b))
  $\wedge$ $(base \in N)$
  $\wedge$ $(1 < base))$
$\rightarrow$ (bign$\Rightarrow$nat (big-plus $(a, b, c, base), base)$
  =   (bign$\Rightarrow$nat $(a, base)$ + bign$\Rightarrow$nat $(b, base)$ + tv$\Rightarrow$nat (c))))

We have proved the theorem above mechanically.

This concludes the formal discussion of the abstract concept of big number addition. We can summarize this section as follows. We defined what a big number (in a given base) is. We defined the natural represented by a given big number. We defined the big number sum of two big numbers. We justified the use of the word ''sum'' by relating big number addition to Peano addition.

4.3. A Piton Program

Our objective is to implement a Piton program that computes 'big-add-array' and 'big-add-carry-out' in the special case in which the input carry flag is **f**. We are not interested in implementing 'big-plus' because our envisioned applications of big number arithmetic use fixed length big numbers. For our purposes, 'big-plus' is simply a mathematical abstraction that is useful in justifying our interest in 'big-add-array' and 'big-add-carry-out'. In Figure 4-1 we show the Piton program named **big-add**. It expects three arguments: **a**, the address of the least significant digit in the first big number array; **b**, the address of the least significant digit in the second big number array; and **n**, the length of the two big number arrays. The base of the big number system is implicitly the first unrepresentable natural on the Piton machine. The subroutine sums the two arrays, overwriting the first big number. It leaves a Piton Boolean on the stack indicating whether the sum "carried out" of the array. More precisely, at the conclusion of the program, the data area addressed by the input value of **a** will contain the successive digits of 'big-add-array' and on top of the temporary stack we will find 'big-add-carry-out'. We will exhibit a formal specification of this program later.

This example illustrates an important aspect of Piton: addresses can be passed as objects. Note that the big numbers to be added are not built into **big-add**. Nor is **big-add** passed the arrays themselves. Instead, pointers to the arrays are passed.

4.4. An Initial State

Most of the work of specifying **big-add** was done when we defined 'bignp', 'big-add-array' and 'big-add-carry-out' and such English phrases as "big number addition." It remains however to cast into a formula the remark that **big-add** "adds the two big numbers together, overwrites the first, and leaves the carry out flag on the stack." This necessarily involves the notions of Piton states, the Piton interpreter, resource errors, etc. Before we embark on this formalization we simply illustrate the behavior of **big-add**—and in so doing familiarize the reader with the structure of Piton states.

To execute **big-add** we will call it from the **main** program shown below. The program assumes that the data segment of our initial p-state contains at least four data areas: arrays **bna** and **bnb** ("Big Number a" and "Big Number b"), each of which is of length n, a global variable **n**, whose value is n, and another global variable **c**. **Main** pushes the starting address of both **bna** and **bnb** onto the stack, pushes their length on the stack, and calls **big-add**. The call will overwrite **bna**. Upon termination of **big-add**, **main** pops the output carry flag off the temporary stack and into the global variable **c** and halts. **Main** has no formals and no temporary variables.

```
(big-add (a b n)                    ; Formal parameters
        nil                         ; Temporary variables
                                    ; Body
    (push-constant (bool f))        ; Push the input carry flag for the
                                    ; first add-nat-with-carry
    (push-local a)                  ; Push the address a

 (dl loop ()                        ; This is the top level loop.
                                    ; Every time we get here the carry
                                    ; flag from the last addition and
                                    ; the current value of a will be
                                    ; on the stack.
    (fetch))                        ; Fetch next digit from a
    (push-local b)                  ; Push the address b
    (fetch)                         ; Fetch next digit from b
    (add-nat-with-carry)            ; Add the two digits and flag
    (push-local a)                  ; Deposit the sum digit in a
    (deposit)                       ; (but leave carry flag)
    (push-local n)                  ; Decrement n by 1
    (sub1-nat)
    (set-local n)                   ; (but leave n on the stack)
    (test-nat-and-jump zero done)   ; If n=0, go to done
    (push-local b)                  ; Increment b by 1
    (push-constant (nat 1))
    (add-addr)
    (pop-local b)
    (push-local a)                  ; Increment a by 1
    (push-constant (nat 1))
    (add-addr)
    (set-local a)                   ; (but leave a on the stack)
    (jump loop)                     ; goto loop
 (dl done ()
    (ret)))                         ; Exit.
```

Figure 4-1: A Piton Program for Big Number Addition

```
(main nil nil
      (push-constant (addr (bna . 0)))
      (push-constant (addr (bnb . 0)))
      (push-global n)
      (call big-add)
      (pop-global c)
      (ret))
```

Suppose we wished to use **main** to solve the big number version of

$$786,433,689,351,873,913,098,236,738$$
$$+ \ \underline{141,915,430,937,733,100,148,932,872}$$
$$?$$

In base 2^{32} (which is 4,294,967,296) these two naturals can be represented by the following big numbers of length 4:

`'((nat 246838082) (nat 3116233281) (nat 42632655) (nat 0))`

and

`'((nat 3579363592) (nat 3979696680) (nat 7693250) (nat 0)).`

A suitable Piton initial state for adding together these two big numbers is shown in Figure 4-2.

The nine fields of the p-state in Figure 4-2 are named in the comments of the figure. We discuss each field in turn. The first is the program counter. Note that it is a tagged address. The **pc** tag indicates that it is an address into the program segment. The pair **(main . 0)** is an address, pointing to the 0th instruction in the **main** program. The second field is the control stack. In this example it contains only one frame and so is of the form **'((**bindings return-pc**))**. Since the current program counter is in **main**, the single frame on the control stack describes the invocation of **main**. Since **main** has no local variables, the frame has the empty list, **nil**, as the local variable bindings. Since there is only one frame on the stack, it describes the top-level entry into Piton and hence the return program counter is completely irrelevant. If control is ever "returned" from this invocation of **main** the Piton machine will halt rather than "return control" outside of Piton. However, despite the fact that the initial return program counter is irrelevant we insist that it be a legal program counter and so in this example we chose **(pc (main . 0))**. The third field is the temporary stack. In this example it is empty. The fourth field is the program segment. It contains two programs, **main** and **big-add**. The fifth field is the data segment. It contains four "global arrays" named, respectively, **bna**, **bnb**, **n** and **c**. **Bna** and **bnb** are both arrays of length four. **N** and **c** are each arrays of length one. We think of **n** and **c** simply as global variables. The **bna** array contains the first of the two big numbers we wish to add, namely **((nat 246838082) (nat 3116233281) (nat 42632655) (nat 0))**. The **bnb** array contains the second big number. **N** contains the (tagged) length of the two arrays. **C** contains the (tagged) natural number 0; the initial value of **c** is irrelevant however. The next three fields are, respectively, the maximum control stack size, **10**, the maximum temporary stack size, **8**, and the word size, **32**. The stack sizes declared in this example are unusually small but sufficient for the computation described. The last field is the program status word **'run**.

Let p_0 be the p-state in Figure 4-2. If one steps this p-state forward 76 times the result is the p-state shown in Figure 4-3. That is, the p-state in Figure 4-3 is equal to $p(p_0, 76)$.

Observe that the psw in Figure 4-3 is **'halt**. This tells us the computation terminated without error. Because the Piton interpreter is a no-op on states with psw **'halt**, we would get the same result had we stepped p_0 more than 76 times. The program counter points to the **ret** statement in the **main** program, the last instruc-

```
p-state('(pc (main . 0)),              ; program counter
       '((nil (pc (main . 0)))),; control stack
      nil,                             ; temporary stack
       '((main nil nil               ; program segment
               (push-constant (addr (bna . 0)))
               (push-constant (addr (bnb . 0)))
               (push-global n)
               (call big-add)
               (pop-global c)
               (ret))
          (big-add (a b n) nil
               (push-constant (bool f))
               (push-local a)
            (dl loop nil (fetch))
               (push-local b)
               (fetch)
               (add-nat-with-carry)
               (push-local a)
               (deposit)
               (push-local n)
               (sub1-nat)
               (set-local n)
               (test-nat-and-jump zero done)
               (push-local b)
               (push-constant (nat 1))
               (add-addr)
               (pop-local b)
               (push-local a)
               (push-constant (nat 1))
               (add-addr)
               (set-local a)
               (jump loop)
            (dl done nil (ret)))),
                                       ; data segment
       '((bna   (nat 246838082) (nat 3116233281)
               (nat 42632655) (nat 0))
          (bnb   (nat 3579363592) (nat 3979696680)
               (nat 7693250) (nat 0))
          (n     (nat 4))
          (c     (nat 0))),
      10, 8,                           ; max ctrl and temp stk sizes
      32, 'run)                        ; word size, psw
```

Figure 4-2: An Initial Piton State for Big Number Addition

```
p-state(' (pc (main . 5)),              ; program counter
     '((nil (pc (main . 0))))),; control stack
    nil,                                ; temporary stack
    '((main nil nil                     ; program segment
            (push-constant (addr (bna . 0)))
            (push-constant (addr (bnb . 0)))
            (push-global n)
            (call big-add)
            (pop-global c)
            (ret))
      (big-add (a b n) nil
            (push-constant (bool f))
            (push-local a)
        (dl loop nil (fetch))
            (push-local b)
            (fetch)
            (add-nat-with-carry)
            (push-local a)
            (deposit)
            (push-local n)
            (sub1-nat)
            (set-local n)
            (test-nat-and-jump zero done)
            (push-local b)
            (push-constant (nat 1))
            (add-addr)
            (pop-local b)
            (push-local a)
            (push-constant (nat 1))
            (add-addr)
            (set-local a)
            (jump loop)
        (dl done nil (ret)))),
                                        ; data segment
     '((bna  (nat 3826201674) (nat 2800962665)
            (nat 50325906) (nat 0))
       (bnb  (nat 3579363592) (nat 3979696680)
            (nat 7693250) (nat 0))
       (n    (nat 4))
       (c    (bool f))),
    10, 8,                              ; max ctrl and temp stk sizes
    32, 'halt)                          ; word size, psw
```

Figure 4-3: A Final Piton State for Big Number Addition

tion executed. The control stack and the temporary stack are exactly as they were in the initial state. The program segment and resource limits are exactly as they were in the initial state—they are never changed. The final value of the **a** array in the data segment is now the big number **((nat 3826201674) (nat 2800962665) (nat 50325906) (nat 0))**. The final value of the global variable **c** is the Boolean value **f**, indicating that the addition did not carry out of the array. A little arithmetic will confirm that the natural represented by the final values of **bna** and **c** is 928,349,120,289,607,013,247,169,610, which is the sum of 786,433,689,351,873,913,098,236,738 and 141,915,430,937,733,100,148,932,872, as desired.

4.5. The Formal Specification

We now develop a formula that expresses the idea that **big-add** computes the big number sum of its two arguments, i.e., overwrites its first argument with the value of 'big-add-array' and pushes the value of 'big-add-carry-out' onto the stack.

4.5.1. Preliminary Definitions

We will need to talk about the arrays associated with given data area names in a given data segment. We will also need to discuss the data segment obtained from another by changing the array associated with a given name. These concepts are easily expressed in terms of functions defined in the formalization of Piton (Appendix II) but because the names used there are unfamiliar we will define slightly more memorable names here.

DEFINITION:
array (*name*, *segment*) = cdr (assoc (*name*, *segment*))

defines the function that returns the array associated with *name* in a given data segment *segment*. 'assoc' is an Nqthm primitive function.

DEFINITION:
put-array (*a*, *name*, *segment*) = put-assoc (*a*, *name*, *segment*)

defines the function that returns a new data segment obtained from *segment* by associating the array *a* with data area name *name* and leaving all other data areas unchanged. 'Put-assoc' is defined on page 239.

So that we can describe the program segment succinctly we will define the constant function 'big-add-program' to be the list constant corresponding to Figure 4-1, page 48.

DEFINITION:
big-add-program
=

```
'(big-add (a b n)          ; Formal parameters
          nil              ; Temporary variables
                           ; Body
```

```
        (push-constant (bool f));   Push the input carry flag for
        ...                               ...
(dl done ()
      (ret)))                        ; Exit.
```

4.5.2. The Initial State

To develop the specification of **big-add** we will consider an "arbitrary" initial p-state in which the current instruction is a legal **call** of **big-add** and we will describe the final p-state produced by executing that **call** statement and all of the **big-add** computation up to and including the return.

The "arbitrary" initial state will be

p-state (pc,
      $ctrl$-stk,
      append (list (tag ('**nat**, n),
                    tag ('**addr**, cons (b, 0)),
                    tag ('**addr**, cons (a, 0))),
          $temp$-stk),
      $prog$-$segment$,
      $data$-$segment$,
      max-$ctrl$-stk-$size$,
      max-$temp$-stk-$size$,
      $word$-$size$,
      '**run**)

where pc is assumed to point to the instruction (**call big-add**) and **big-add** is defined as in Figure 4-1. Let p_0 be the p-state above. Then the additional constraints mentioned can be formalized by saying p-current-instruction (p_0) = '(**call big-add**) and definition ('**big-add**, $prog$-$segment$) = big-add-program.

Observe that the psw of p_0 is '**run**. Note also that the temporary stack in p_0 consists of some arbitrary $temp$-stk with three additional items pushed onto it. The items (in the order in which they were pushed) are a tagged data address to location 0 of the data area named a, a tagged data address to location 0 of the data area named b, and a tagged natural n. Note carefully that the a, b, and n used here are variable symbols which are so far unconstrained. In summary, p_0 is an arbitrary p-state poised to execute a **call** of our **big-add** on three arguments that are tagged in accordance with our expectations. However, much more needs to said about those arguments.

4.5.3. The Preconditions

The "expected" values of a and b are the names of data areas that contain big numbers of equal length and the "expected" value of n is the length of those big numbers. These "expectations" are part of the "input conditions" of **big-add** (which heretofore have been implicit in our discussions) and the formula

$\quad$ definedp $(a,\text{p-data-segment}\,(p_0))$

$\wedge\quad$ definedp $(b,\text{p-data-segment}\,(p_0))$

$\wedge\quad (a \neq b)$

$\wedge\quad$ bignp $(\text{array}\,(a,\text{p-data-segment}\,(p_0)),\ 2^{\text{p-word-size}\,(p_0)})$

$\wedge\quad$ bignp $(\text{array}\,(b,\text{p-data-segment}\,(p_0)),\ 2^{\text{p-word-size}\,(p_0)})$

$\wedge\quad (n = \text{length}\,(\text{array}\,(a,\text{p-data-segment}\,(p_0))))$

$\wedge\quad (n = \text{length}\,(\text{array}\,(b,\text{p-data-segment}\,(p_0))))$

makes them explicit and formal.

The function 'p-data-segment' is one of the accessors of the 'p-state' shell constructor and returns the data segment of the state (see page 218). The function 'definedp' checks that its first argument is the name of a data area in its second argument and is defined on page 180. Observe that we explicitly assume that 'a' is different from 'b', i.e., that the program is operating on two distinct data areas. (Of course, they may contain the same big number.) This is not technically necessary; **big-add** performs in a meaningful way even if it is passed the same address in both arguments. But by ruling out this possibility we simplify our analysis of the program somewhat.

The above conditions are the obvious preconditions for **big-add**. However, if **big-add** is to run without error there are several more details we must consider. We address them in roughly the order in which they arise in the execution of the code for **big-add**.

In order for the **call** statement to execute without error, we must know that there is enough room on the control stack to build the new frame. Inspection of the definition of 'p-ctrl-stk-size' (page 191) shows that the frame we will build has size five (three for the local variables of **big-add** plus two more). Thus, we must assume p-max-ctrl-stk-size $(p_0) \geq (5 + \text{p-ctrl-stk-size}\,(\text{p-ctrl-stk}\,(p_0)))$.

We must similarly worry about the temporary stack overflowing during the **big-add** computation. Once we enter **big-add** the temporary stack will be *temp-stk* (because the three actuals will have been popped off). By how far will we extend *temp-stk* during the computation? Inspection will show that we need at most three more slots. For example, once we have executed the first **(push-local b)** we have pushed three items onto the stack. At no point do we have more than three items pushed. It is just a coincidence that the amount of temporary stack needed by **big-add**'s body is the same as the number of actuals on the stack at the time of the **call**. In any case, we must know p-max-temp-stk-size $(p_0) \geq (3 + \text{length}\,(\text{p-temp-stk}\,(p_0)))$. If this assumption is violated, some push in the body of **big-add** causes a stack overflow error.

Once we enter the **loop** in **big-add** we fetch the first digit from its local variable **a**. But this assumes there is a first digit, i.e., that the length, n, of the argument arrays is nonzero. **Big-add** could have been coded to work for empty big numbers, by checking **n** before the first **fetch**. But as it is coded, **big-add** assumes the big numbers are nonempty and does not check **n** until it has added the low order digits together and decremented **n**. Thus, we must assume $\neg\,(n \cong 0)$. If this assumption is violated, **big-add** causes an addressing error.

When we decrement **n** with **sub1-nat** we must know that n is a representable natural. Thus, $n < 2^{\text{p-word-size}\,(p_0)}$. If this assumption is violated, **big-add** causes an arithmetic error. We can prove from the foregoing assumptions that we will not get addressing errors when we increment **a** and **b** n times.

Finally, when we execute the **ret** statement at the conclusion of **big-add** we must know that the initial control stack was nonempty. Thus, listp (p-ctrl-stk (p_0)). No Piton state should have an empty control stack.

For convenience, we collect all of these conditions together into a single predicate.

DEFINITION:
big-add-input-conditionp (a, b, n, p_0)

$=$

$\quad$ definedp $(a,\ \text{p-data-segment}\,(p_0))$

$\wedge\quad$ definedp $(b,\ \text{p-data-segment}\,(p_0))$

$\wedge\quad (a \neq b)$

$\wedge\quad$ bignp $(\text{array}\,(a,\ \text{p-data-segment}\,(p_0)),\ 2^{\text{p-word-size}\,(p_0)})$

$\wedge\quad$ bignp $(\text{array}\,(b,\ \text{p-data-segment}\,(p_0)),\ 2^{\text{p-word-size}\,(p_0)})$

$\wedge\quad (n = \text{length}\,(\text{array}\,(a,\ \text{p-data-segment}\,(p_0))))$

$\wedge\quad (n = \text{length}\,(\text{array}\,(b,\ \text{p-data-segment}\,(p_0))))$

$\wedge\quad (\text{p-max-ctrl-stk-size}\,(p_0) \geq (5 + \text{p-ctrl-stk-size}\,(\text{p-ctrl-stk}\,(p_0))))$

$\wedge\quad (\text{p-max-temp-stk-size}\,(p_0) \geq (3 + \text{length}\,(\text{p-temp-stk}\,(p_0))))$

$\wedge\quad (\neg\,(n \cong 0))$

$\wedge\quad (n < 2^{\text{p-word-size}\,(p_0)})$

$\wedge\quad$ listp $(\text{p-ctrl-stk}\,(p_0))$

In our correctness theorem we will assume that big-add-input-conditionp (a, b, n, p_0) holds.

4.5.4. The Final State

Next we wish to characterize the state obtained by executing the above **call** of **big-add**. Because of the constructive nature of the Nqthm logic, it will be necessary to say how many instructions we wish to execute.[6] For the moment though, let *clock* stand for some expression that determines the number of Piton instructions

---

[6]This is not strictly true. It is possible to phrase partial correctness theorems by including among the hypotheses the assumption that k is a number such that the psw of p (p_0, k) is **'halt**. We do not pursue this here.

executed from the **call** above to the **ret** instruction at the end of that invocation of **big-add**, inclusive. We will define *clock* in the next section. Then the final state obtained by executing *clock* instructions starting at p_0 is $p(p_0, clock)$. We would like to characterize $p(p_0, clock)$ completely.

The program counter of the final state will be one greater than *pc*. The control stack will be exactly the control stack of p_0. The temporary stack will be *temp-stk* with

big-add-carry-out (array $(a, data\text{-}segment)$, array $(b, data\text{-}segment)$, **f**, $2^{word\text{-}size}$)

pushed on top of it. The program segment of the final state will be the same as the program segment of p_0. The data segment of the final state will be the same as the data segment of p_0 with one exception:

big-add-array (array $(a, data\text{-}segment)$, array $(b, data\text{-}segment)$, **f**, $2^{word\text{-}size}$)

will be the array associated with the data area named a. The resource limitations of the final state will be the same as those of the initial state and the psw will still be **'run**. Thus, $p(p_0, clock)$ is equal to

p-state (add1-addr (pc),
  ctrl-stk,
  push (big-add-carry-out (array $(a, data\text{-}segment)$,
                array $(b, data\text{-}segment)$,
                f,
                $2^{word\text{-}size}$),
        temp-stk),
  prog-segment,
  put-array (big-add-array (array $(a, data\text{-}segment)$,
                array $(b, data\text{-}segment)$,
                f,
                $2^{word\text{-}size}$),
        a,
        data-segment),
  max-ctrl-stk-size,
  max-temp-stk-size,
  word-size,
  'run)

The function 'add1-addr' is defined on page 178 as part of the formalization of Piton.

4.5.5. The Clock

It remains only to say what *clock* is. We define a function, 'big-add-clock', which determines the number of Piton instructions necessary to execute any legal **call** of **big-add**. In the case of **big-add** this function is a function only of the length, n, of the big numbers being added. The **call** itself costs one instruction. Inspection of the code for **big-add** (page 48) shows that we then execute two more instructions before arriving at the label **loop**. Suppose that 'big-add-loop-clock' is defined to be the number of instructions it takes to complete the loop and return. Then

DEFINITION:
big-add-clock $(n) = $ **3** $+ $ big-add-loop-clock (n)

To define 'big-add-loop-clock' we walk through the code symbolically again. If the value of **n** is 1 we execute 11 instructions from **loop** through the **ret** at **done**. If the value of **n** is not 1, we execute 19 instructions and arrive back at **loop** with **n** decremented by 1. Thus, a suitable definition of 'big-add-loop-clock' is:

DEFINITION:
big-add-loop-clock (n)

=

if $n \cong $ **0 then 0**
elseif $n = $ **1 then 11**
else 19 $+ $ big-add-loop-clock $(n -1)$ **endif**

The case in which n is 0 never arises but is included in the above definition to insure that the function defined is total.   Of course, it is easy to see that big-add-loop-clock (n) is **11** $+ ($ **19** $\times (n -1))$, when n is non-0.   While such an algebraic expression of the clock is pleasing, we prefer the recursive formulation in general because it mirrors the exploration of the code and more easily accommodates special cases (e.g., interior branches).

It is not strictly necessary to characterize exactly the number of instructions required to execute a program.   One can state correctness results conditionally on the termination of the program (in the sense of including a hypothesis about the variable n being such that the psw of p (p_0, n) is **'HALT**) and state partial correctness results. But since the Nqthm logic does not provide existential quantification we cannot say "there exists an n such that the program terminates;" the nearest we can come is to exhibit constructively such an n.   In that sense, 'big-add-clock' is a "witness function" for an implicit existential quantification. We discuss this issue further on page 86.

4.5.6. The Correctness Theorem

The correctness theorem for **big-add** can now be written down completely.   It is shown in Figure 4-4 and should be self-explanatory.   The formula is a theorem.   It can be proved from the foregoing definitions and the formal definition of Piton.   In fact, we have proved it mechanically.   We discuss the proof later (see page 65).

The theorem of Figure 4-4 is very powerful.   It can be applied to any legal call of **big-add**, no matter what other programs are in the program segment and no matter what data areas are defined in the data segment.   It specifies exactly how many instructions will be executed on behalf of the call.   For example, to add together two big numbers of length 4 will take 71 Piton instructions.   To add together two big numbers of length 100 will take 1,895 Piton instructions.   The theorem tells us exactly how to obtain the final state: pop the arguments off the stack, push the output carry on the stack, deposit the big number sum in the data area of the first argument, and keep the psw **'run**.   Note that since we know the final psw is **'run** we know that no run-time errors occur if the preconditions are satisfied.   The beauty of the theorem in Figure 4-4 is that for all intents and purposes it allows us to treat **(call big-add)** as a Piton primitive.   We illustrate this in the next section.

THEOREM: Correctness of **big-add**
((p_0
  = p-state (pc,
            ctrl-stk,
            append (list (tag ('**nat**, n),
                         tag ('**addr**, cons (b, **0**)),
                         tag ('**addr**, cons (a, **0**))),
                    temp-stk),
            prog-segment,
            data-segment,
            max-ctrl-stk-size,
            max-temp-stk-size,
            word-size,
            '**run**))
  ∧ (p-current-instruction (p_0) = ' (**call big-add**))
  ∧ (definition ('**big-add**, prog-segment) = big-add-program)
  ∧ big-add-input-conditionp (a, b, n, p_0))
→ (p (p_0, big-add-clock (n))
  = p-state (add1-addr (pc),
            ctrl-stk,
            push (big-add-carry-out (array (a, data-segment),
                                     array (b, data-segment),
                                     f,
                                     $2^{word-size}$),
                  temp-stk),
            prog-segment,
            put-array (big-add-array (array (a, data-segment),
                                      array (b, data-segment),
                                      f,
                                      $2^{word-size}$),
                       a,
                       data-segment),
            max-ctrl-stk-size,
            max-temp-stk-size,
            word-size,
            '**run**))

Figure 4-4: The Specification of **big-add**

The theorem stated in Figure 4-4 could be packaged more compactly. For example, after suitably defining some functions we could state the theorem as

THEOREM:
   big-add-hyp (a, b, n, p_0)
→ (p (p_0, big-add-clock (n)) = big-add-next-state (a, b, n, p_0))

and read it as "if p_0 is a Piton state poised to execute **big-add** on the big number arrays in data areas a and b (each of which is of length n), then stepping p_0 forward big-add-clock (n) steps produces the desired next state.'' Some readers may find the smaller statement more appealing, even though it involves the definition of symbols used nowhere else.

4.6. Using the Formal Specification

In this section we discuss how to ''stack'' correctness proofs for Piton programs. We explain how the above theorem about **big-add** can be used to construct a correctness proof for a program that uses **big-add**. This may help some readers understand why we chose the above form for our correctness theorem. In addition, it suggests that we can verify systems of Piton programs by verifying the individual subroutines—which of course we can but which we have not yet done for systems of nontrivial complexity. We here define a trivial two-layered ''system,'' consisting simply of **big-add** and a top-level main program which uses **big-add**, together with a particular data segment. We prove the correctness of that system, assuming the above theorem about **big-add**. The style of our proof illustrates how to go about proving the correctness of any program (top-level or not) which uses **big-add**.

Recall the **main** program on page 48 which uses **big-add** to add **bna** and **bnb**. Define the constant function '[main-program]' to be equal to the list constant describing **main**.

DEFINITION:
main-program
=
```
'(main nil nil
  (push-constant (addr (bna . 0)))
  (push-constant (addr (bnb . 0)))
  (push-global n)
  (call big-add)
  (pop-global c)
  (ret))
```

Our earlier use of **main** was in Figure 4-2 (page 50) where it was the top-level program in a p-state configured to add together two specific big numbers.

Now consider the function

DEFINITION:
system-initial-state (a, b)

=

p-state (`'(pc (main . 0))`,
      `'((nil (pc (main . 0))))`,
      nil,
      list (main-program, big-add-program),
      list (cons (`'bna`, a),
          cons (`'bnb`, b),
          cons (`'n`, list (tag (`'nat`, length (a)))),
          cons (`'c`, list (tag (`'nat`, 0)))),
      10,
      8,
      32,
      `'run`)

If given two big numbers (base 2^{32}) of equal length, this function creates an initial p-state suitable for adding them together and storing the answers in **bna** and **c**. The p-state shown in Figure 4-2 was actually created by applying 'system-initial-state' to the particular big numbers used in the earlier example. We would like to prove the correctness of the system of Piton programs and data described by 'system-initial-state'.

We claim that the state produced by 'system-initial-state' is suitable for adding big numbers together provided a and b are both nonempty big numbers (base 2^{32}) of length n, where n is less than 2^{32}. We define the following predicate to check these conditions.

DEFINITION:
system-initial-state-okp (a, b)

=

$\quad$ bignp $(a, 2^{32})$
$\wedge \quad$ bignp $(b, 2^{32})$
$\wedge \quad$ (length (a) = length (b))
$\wedge \quad$ ($\neg$ (length $(a) \cong 0$))
$\wedge \quad$ (length $(a) < 2^{32}$)

How long does it take for this system to run to completion? The answer is provided by the function

DEFINITION:
system-initial-state-clock (a, b) = 5 + big-add-clock (length (a))

That is, **main** executes five instructions in addition to the **call** of **big-add**.

The following formula describes the correctness of the system:

THEOREM:
    system-initial-state-okp (a, b)
$\rightarrow$ (p (system-initial-state (a, b), system-initial-state-clock (a, b)))
  = p-state(`'(pc (main . 5))`,
              `'((nil (pc (main . 0))))`,
              nil,
              list (main-program, big-add-program),
              list (cons (`'bna`, big-add-array $(a, b, \mathbf{f}, 2^{32})$),
                    cons (`'bnb`, b),
                    cons (`'n`, list (tag (`'nat`, length (a))))),
                    cons (`'c`, list (big-add-carry-out $(a, b, \mathbf{f}, 2^{32})$)))),
              10,
              8,
              32,
              `'halt`))

That is, if a and b satisfy 'system-initial-state-okp' then the result of running the state system-initial-state (a, b) for system-initial-state-clock (a, b) instructions is a **halt**ed p-state in which the array named **bna** has the value big-add-array $(a, b, \mathbf{f}, 2^{32})$ and the global variable **c** has the value big-add-carry-out $(a, b, \mathbf{f}, 2^{32})$.

To prove this we must first observe an important theorem about the Piton interpreter 'p'.

THEOREM: Sequential Execution
$$p(s, i+j) = p(p(s, i), j)$$

This theorem says that running forward $i + j$ steps is the same as running forward i steps and then running forward j steps from there. The proof is trivial by induction on s and i.

We now present the proof of the correctness of our **big-add** system. Readers interested in constructing the proof formally should first read Appendix II, where we present the formal definition of Piton. Because this is the first Piton proof we have discussed, we will go very slowly. The proof is in fact immediate from the correctness of **big-add**.

Proof. We will prove the correctness of our **big-add** system by reducing the left-hand side of the conclusion, which we'll call p_{left}, to the right-hand side, which we'll call p_{right}. Let p_0 be system-initial-state (a, b). Then p_{left} is p $(p_0,$ system-initial-state-clock $(a, b))$. But system-initial-state-clock (a, b) is **5** + big-add-clock (length (a)), which can be written as **3** + (big-add-clock (length (a)) + **2**). Thus, by the sequential execution theorem, p_{left} can be equivalently obtained by composing three smaller runs of 'p'. The first run, which we will call *Run 1* starts from p_0, takes three steps, and produces a state we'll call p_3. The second run, *Run 2*, starts at p_3, takes big-add-clock (length (a)) steps, and produces a state we'll call p_{3+c}. The third run, *Run 3*, starts at p_{3+c}, takes two steps, and produces a state we'll call p_{done}. By the sequential execution lemma, p_{left} is equal to p_{done}. It will turn out that p_{done} is identical to p_{right}, as desired. We proceed by working on each of the three runs in turn.

Run 1. Our starting state, p_0, is system-initial-state (a, b), which is

p-state(`'(pc (main . 0))`,

```
'((nil (pc (main . 0)))),
nil,
```
list (main-program, big-add-program),
list (cons (**'bna**, *a*),
    cons (**'bnb**, *b*),
    cons (**'n**, list (tag (**'nat**, length (*a*)))),
    cons (**'c**, list (tag (**'nat**, 0)))),
```
10,
8,
32,
'run)
```

by the definition of 'system-initial-state'. We wish to obtain p_3, which is p(p_0, 3). Recall how 'p' is defined: if the precondition of the current instruction is satisfied, the next state is that obtained by applying the current step function to the old state. The program counter of p_0 points to the 0th instruction of **main**. Thus, the current instruction is **(push-constant (addr (bna . 0)))**. The preconditions of **push-constant** are satisfied: there is room to push at least one item on the temporary stack because the maximum stack length is 8 and the stack is currently empty. Thus, p(p_0, 3) is equal to p(p_1, 2), where p_1 is the result of stepping p_0 with the **push-constant** step function. The **push-constant** step function increments the program counter by one and pushes the given constant onto the temporary stack. Thus, p_1 is

p-state (**'(pc (main . 1))**,
    '((nil (pc (main . 0)))),
    '((addr (bna . 0))),
    list (main-program, big-add-program),
    list (cons (**'bna**, *a*),
        cons (**'bnb**, *b*),
        cons (**'n**, list (tag (**'nat**, length (*a*)))),
        cons (**'c**, list (tag (**'nat**, 0)))),
```
    10,
    8,
    32,
    'run)
```

The derivation of p_1 from p_0 illustrates the most fundamental Piton proof technique: determine the current instruction, check that the precondition is satisfied, apply the step function, and simplify. This technique is called *symbolic execution* of Piton.

We can compute p(p_1, 2), by two more applications of symbolic execution, first running the **push-constant** instruction at **(pc (main . 1))** (which increments the program counter and pushes the **bnb** address) and then running the **push-global** instruction at **(pc (main . 2))** (which increments the program counter and pushes the current value of **n**). The result is

p-state (**'(pc (main . 3))**,
    '((nil (pc (main . 0)))),
    list (tag (**'nat**, length (*a*)),

```
        '(addr (bnb . 0)),
        '(addr (bna . 0))),
  list (main-program, big-add-program),
  list (cons ('bna, a),
        cons ('bnb, b),
        cons ('n, list (tag ('nat, length (a)))),
        cons ('c, list (tag ('nat, 0)))),
  10,
  8,
  32,
  'run)
```

which we are calling p_3.

Run 2. We now wish to obtain p_{3+c}, which by definition is $p(p_3,$ big-add-clock (length (a))). Observe that the current instruction of p_3 is (**call big-add**). This can be deduced from the fact that the 'p-pc' above is '(**pc (main . 3)**) and the instruction (**call big-add**) is at position **3** in the definition of '**main** in the 'p-prog-segment'. If we used symbolic execution here we would next investigate the precondition for **call** and apply the **call** step function, taking us into the body of **big-add**. That would be a strategic mistake. Instead, we will appeal to the correctness theorem for **big-add**.

First, observe that the temporary stack of p_3 can be equivalently written as an 'append' of three **tagged** objects to the empty stack. That is,

```
p-state ('(pc (main . 3)),
         '((nil (pc (main . 0)))),
         append (list (tag ('nat, length (a)),
                       tag ('addr, '(bnb . 0)),
                       tag ('addr, '(bna . 0))),
                 nil),
         list (main-program, big-add-program),
         list (cons ('bna, a),
               cons ('bnb, b),
               cons ('n, list (tag ('nat, length (a)))),
               cons ('c, list (tag ('nat, 0)))),
         10,
         8,
         32,
         'run)
```

is just another way to write p_3. But now we see we are in exactly the situation handled by the theorem assumed in the previous section (see page 58), the correctness of **big-add**: we have a state whose current instruction is a call to our **big-add** program on legal input and we wish to run forward big-add-clock (length (a)) steps.

Thus, instantiating the right-hand side of the conclusion of the correctness of **big-add**,

```
p-state ('(pc (main . 4)),
         '((nil (pc (main . 0)))),
```

```
      push (big-add-carry-out (a, b, f, 2³²), nil),
      list (main-program, big-add-program),
      list (cons ('bna, big-add-array (a, b, f, 2³²)),
           cons ('bnb, b),
           cons ('n, list (tag ('nat, length (a)))),
           cons ('c, list (tag ('nat, 0)))),
      10,
      8,
      32,
      'run)
```

gives us p_{3+c}. This concludes the second run.

Run 3. We obtain p_{done} by stepping forward from p_{3+c} two more steps. We use symbolic execution again, running the **pop-global** at **(pc (main . 4))** and then the **ret** at **(pc (main . 5))**. The result is

```
p-state (' (pc (main . 5)),
      ' ((nil (pc (main . 0)))),
      nil,
      list (main-program, big-add-program),
      list (cons ('bna, big-add-array (a, b, f, 2³²)),
           cons ('bnb, b),
           cons ('n, list (tag ('nat, length (a)))),
           cons ('c, big-add-carry-out (a, b, f, 2³²))),
      10,
      8,
      32,
      'halt)
```

which is the desired final state, **p$_{right}$**. **Q.E.D.**

The point of this exercise is that the correctness theorem for **big-add** permitted us to prove **main** correct without ever considering the code for **big-add**. Observe that all of the program counters in the states mentioned in the proof above are in **main**. Observe also that our correctness result for **big-add** readily applied to a state containing another program (i.e., **main**) and data areas irrelevant to **big-add** (i.e., **c**). Finally, observe that the sequential execution theorem freed us from having to consider exactly how many clock ticks were available when we encountered the call of **big-add**. If there are at least big-add-clock (n), where n is the length of the big numbers in question, then we can step past the **call** and decrement the clock by big-add-clock (n). Note this is one advantage of having counted precisely the number of instructions required. In essence, the correctness theorem for **big-add** permits us to treat **(call big-add)** as though it were a primitive instruction; in one proof step we obtain the state produced by any successful execution.

4.7. The Proof of the Correctness of Big-Add

We now back up and sketch the proof of the correctness of the **big-add** program, which was used in the proof just discussed.

As seen in the proof above, it is relatively straightforward to deal with "straight line" Piton code. One merely symbolically executes the definition of Piton, accumulating into the state the successive changes. **Big-add** is more subtle because it has a loop in it. The presence of the loop immediately suggests induction, which in turn suggests that we need some more general concepts with which to discuss the "invariants" maintained at intermediate arrivals at the top of the loop. So much for generalities.

The key to our proof of the correctness of **big-add** is to define two functions derived from the loop in **big-add** that compute the final big number array and the final carry out "in the same way" that the Piton loop does. We call these derived functions 'big-add-array-loop' and 'big-add-carry-out-loop'. We then prove that the derived functions are (a) computed by the loop in **big-add** and (b) are equivalent to those used in the abstract specification.

4.7.1. The Derived Specification Functions

The following function can be thought of as the essence of the loop in **big-add** *vis-a-vis* its effect on the array stored in the data area named *a* in the correctness theorem for **big-add**. Let *a-array* and *b-array* be the arrays associated with the *a* and *b* data areas of the correctness theorem. The variable *n*, below, plays the same role below as **n** does in **big-add**, namely, it is the distance from the current location, *i*, to the end of the arrays. That is, *n* is the number of iterations we are to perform. *C* below is the truth value corresponding to the current input carry flag and *base* is the base of the big number system.

DEFINITION:
big-add-array-loop $(i, a\text{-}array, b\text{-}array, n, c, base)$

=

if $(n-1) \cong 0$
 then put (tag (**'nat**,

                  untag (get $(i, a\text{-}array)$)) + untag (get $(i, b\text{-}array)$)) + tv$\Rightarrow$nat (c)

            mod $base$),

      i,

     $a\text{-}array$)

 else big-add-array-loop $(1+i,$

                put (tag (**'nat**,

                        untag (get $(i, a\text{-}array)$)

                  +   untag (get $(i, b\text{-}array)$)

                  +   tv$\Rightarrow$nat (c)

              mod $base$),

            i,

           $a\text{-}array$),

         $b\text{-}array$,

        $n-1$,

         (untag (get $(i, a\text{-}array)$)

        +   untag (get $(i, b\text{-}array)$)

        +   tv$\Rightarrow$nat (c))

      $\geq$ $base$,

      $base$) **endif**

The function 'put' is defined on page 239 as part of the definition of Piton. put (val, i, a) puts val at the ith location of the array a and returns the resulting array. Observe that 'big-add-array-loop' iterates n times, considering successive array positions starting at position i, and 'put's into each position the corresponding digit of the big number sum of the two arrays.

We also define the analogous derived function which describes the final carry out computed by the loop and call it 'big-add-carry-out-loop'. We do not exhibit its definition here.

These derived functions let us separate the problem of what the Piton program computes from the problem of whether it computes the specified quantities. We discuss these two problems in the next two sections.

4.7.2. The Equivalence of the Derived Functions and the Piton Code

The next step in our proof is to establish that the derived functions indeed describe the effects of the loop in **big-add**. This immediately raises another Piton specification challenge.

In Figure 4-5 we specify the loop in **big-add**. The formula, which is very reminiscent of the specification of **big-add** itself, is of the form "$hyp \rightarrow p(p_0, k) = p_k$," where p_0 is the initial state of a Piton computation of length k and p_k is the final state. In this theorem, hyp restricts the initial state to an "arbitrary" legal arrival at **loop** in **big-add** and k measures a run up through the execution of the **ret**

THEOREM:

((length (array (a, *data-segment*)) < $2^{word\text{-}size}$)

$\wedge$ ($\neg$ (*word-size* $\cong$ 0))

$\wedge$ listp (*ctrl-stk*)

$\wedge$ bignp (array (a, *data-segment*), $2^{word\text{-}size}$)

$\wedge$ bignp (array (b, *data-segment*), $2^{word\text{-}size}$)

$\wedge$ (*max-temp-stk-size* $\geq$ (1+ (1+ (1+ length (*temp-stk*)))))

$\wedge$ (definition ('**big-add**, *prog-segment*) = big-add-program)

$\wedge$ definedp (a, *data-segment*)

$\wedge$ definedp (b, *data-segment*)

$\wedge$ ($a \neq b$)

$\wedge$ ($i \in$ N)

$\wedge$ ($i < n$)

$\wedge$ (n = length (array (a, *data-segment*)))

$\wedge$ (n = length (array (b, *data-segment*)))

$\wedge$ booleanp (c))

$\rightarrow$ (p (p-state ('(**pc** (**big-add** . 2)),
           push (p-frame (list (cons ('**a**, tag ('**addr**, cons (a, i))),
                           cons ('**b**, tag ('**addr**, cons (b, i))),
                           cons ('**n**, tag ('**nat**, $n - i$))),
                    ret-pc),
            ctrl-stk),
           push (tag ('**addr**, cons (a, i)), push (tag ('**bool**, c), *temp-stk*)),
           prog-segment,
           data-segment,
           max-ctrl-stk-size,
           max-temp-stk-size,
           word-size,
           '**run**),
       big-add-loop-clock ($n - i$))

=   p-state (*ret-pc*,
        ctrl-stk,
        push (big-add-carry-out-loop (i,
                            array (a, *data-segment*),
                            array (b, *data-segment*),
                            $n - i$,
                            tv (c),
                            $2^{word\text{-}size}$),
           temp-stk),
        prog-segment,
        put-array (big-add-array-loop (i,
                          array (a, *data-segment*),
                          array (b, *data-segment*),
                          $n - i$,
                          tv (c),
                          $2^{word\text{-}size}$),
              a,
              data-segment),
        max-ctrl-stk-size,
        max-temp-stk-size,
        word-size,
        '**run**))

Figure 4-5: The Specification of the Loop in **big-add**

statement in **big-add**. Note that the hypothesis is basically that part of 'big-add-input-conditionp' that is needed to get us through the loop without error. The program counter in the initial state is **(pc (big-add . 2))**, which points to the instruction labeled **loop**. The "shape" of the control stack in the initial state is as constructed by any legal call of **big-add**, that is, the top frame contains bindings for the three locals **a**, **b**, and **n**, and the usual return program counter. The values of the three locals are consistent with the idea that we are considering an arbitrary legal arrival at **loop**, not just the first arrival. That is, **a** and **b** point to the ith words of their respective data areas, a and b, and **n** has been decremented by i from its original value of length (a). The temporary stack is also configured as though we were at an arbitrary arrival at **loop**. We see two things pushed onto it, a running carry and the current value of **a**. Finally, the instruction count controlling how many instructions we execute is big-add-loop-clock $(n - i)$, just enough instructions to take us right through the loop and out via the **ret** instruction.

The final state described by the theorem in Figure 4-5 is similar to that for **big-add** except that we used the derived functions 'big-add-carry-out-loop' and 'big-add-array-loop'. Furthermore, it is consistent with the idea that we started at an arbitrary arrival at **loop**. Thus, the carry flag on the stack is that obtained by starting at position i and iterating $n - i$ times and the final value of **a** is the analogous big number array computed from position i onwards.

Because we used 'big-add-carry-out-loop' and 'big-add-array-loop' in this specification it is relatively straightforward to prove this theorem. The induction is on i up to n. In the base case, when the difference between n and i is 1, the proof requires the symbolic execution of 11 Piton instructions (from **loop** through the **ret** statement) to reduce the left-hand side to the right-hand side. In the inductive step, the proof requires the symbolic execution of 19 instructions (once around the **loop**) to reduce the induction conclusion to the induction hypothesis. No deep (big number specific) problems arise in the proof since 'big-add-carry-out-loop' and 'big-add-array-loop' do exactly the same sequence of 'put's as the code. Basically this entire proof is devoted merely to checking that all the preconditions of all the Piton instructions are satisfied under the hypotheses listed and that the step functions compose as claimed by the derived functions. We do not discuss the proof of the theorem further. Instead, we turn to the proof of the correctness of **big-add** given this theorem about its **loop**.

Recall that the correctness of **big-add** describes an initial state in which we are about to **call big-add** on arguments satisfying the input condition of **big-add**. We are asked to execute **3** + big-add-loop-clock (n) instructions. The first instruction is a **call** and we build a standard **big-add** frame with the initial values of **a**, **b**, and **n**. In that frame, **a** and **b** both point to the 0th position of their respective arrays. We then enter the body of **big-add** and execute two more instructions. These push two things onto the temporary stack, an initial input carry of **f** and the value of **a**. We then arrive at **loop** with big-add-loop-clock (n) ticks left on the clock. The theorem of Figure 4-5 applies if we let i be 0 and c be **f**. We therefore conclude that the final state is the corresponding instance of the final p-state of Figure 4-5. Note that this instance is exactly the state described in the correctness of **big-add** except that in the instance at hand we see the derived functions instead of the original specification functions. That is, the proof would be finished if we just knew that big-add-carry-out-loop (0, a, b, length(a), f, $base$) were the same as big-add-carry-out (a, b, f, $base$) and that big-add-array-loop (0, a, b, length(a), f, $base$) were the same as big-add-array (a, b, f, $base$).

4.7.3. The Equivalence of the Derived and Original Functions

To complete the proof it remains only to show the equivalence of the derived and original specification functions. We will focus on the 'big-add-array' versus 'big-add-array-loop' and leave the carry out case for the reader.

We wish to prove

THEOREM:
$$(\text{listp}(a) \wedge \text{bignp}(a, base) \wedge (base \in N) \wedge (1 < base))$$
$$\rightarrow \quad (\text{big-add-array}(a, b, f, base)$$
$$= \text{big-add-array-loop}(0, a, b, \text{length}(a), f, base))$$

which asserts that if 'big-add-array-loop' iterates length(a) times starting from position 0 the result is the same big number as that produced by 'big-add-array'.

To prove the theorem above we formulate a more general lemma that can be proved by induction. One hint that this is necessary is that one must inductively unfold 'big-add-array-loop' but when the first argument is 0 it is impossible to have an inductive instance of the theorem with the first argument being one greater. We therefore prove

LEMMA:
$$((i \in N) \wedge (i < \text{length}(a)) \wedge \text{bignp}(a, base) \wedge (base \in N) \wedge (1 < base))$$
$$\rightarrow (\quad \text{big-add-array}(\text{nth-cdr}(i, a), \text{nth-cdr}(i, b), c, base)$$
$$= \quad \text{nth-cdr}(i, \text{big-add-array-loop}(i, a, b, \text{length}(a) - i, c, base)))$$

where nth-cdr (i, x) is defined so that it is the ith 'cdr' of x. The lemma above says that if you use 'big-add-array' to get the sum of the ith 'cdr's of a and b you get the same thing as using 'big-add-array-loop' on all of a and b but starting at the ith position and iterating length(a) − i times and then looking at the ith 'cdr' of the result. Note that we also generalized the input carry flag from f to any c. The theorem can be proved by induction on i up to length(a). The proof is not entirely trivial. For example, we here deal with the question of whether the computation is "messed up" by the repeated "destruction" of **a**. That is, each time 'big-add-array-loop' iterates the a it uses subsequently is altered by a 'put' at the location just inspected. "Fortunately" the alteration occurs to the left of that portion of a that the subsequent computation inspects.

The theorem that 'big-add-array' is 'big-add-array-loop' follows immediately from the lemma by instantiating i above with 0 and c with f and then simplifying nth-cdr $(0, x)$ to x and length(a) − 0 to length(a).

The corresponding equivalence theorem for the carry out case is similar. Once both of these equivalence theorems are proved the proof of the correctness of **big-add** follows immediately.

4.8. Summary

We can summarize this chapter as follows. We formally defined big number addition. We showed a Piton program for doing big number addition and built a simple system that called the program. We illustrated Piton states by setting up an addition problem and running it on the Piton machine. We illustrated how Piton programs can be formally specified by theorems in the logic. We showed how specifications in the style proposed can be used, together with symbolic execution, to prove the correctness of Piton programs and systems.

This concludes our discussion of **big-add**. Please recall that this chapter was essentially a detour intended to familiarize the reader with Piton and with how it is possible to specify and prove the correctness of Piton programs from the formal semantics of Piton. However, this book is not about big number arithmetic, how to prove Piton programs correct or even how to program in Piton. It is about how we implemented Piton on the FM9001, how we specified the correctness of that implementation, and how we proved it. Now that the semantics of Piton are clearer, we return to the mainstream of this book.

5

A Sketch of FM9001

We wish to implement Piton on FM9001. In this chapter we explain how FM9001 works and the sense in which it has been implemented correctly in hardware. This chapter is necessarily only a sketch of a large body of work by Brock and Hunt and reported by them in [7].

5.1. FM9001 Fetch-Execute Cycle

The FM9001 is a 32-bit general purpose microprocessor. The machine provides 16 general purpose 32-bit wide registers, numbered 0 through 15, four 1-bit condition code registers, designated "carry," "zero," "overflow," and "negative," and up to 2^{32} words of memory. Register 15 is used as the program counter.

The instruction formats and semantics are summarized by Figure 5-1, which was taken, with permission, from [7]. Very informally, and inaccurately, each instruction computes a specified operation on two specified operands, A and B, and stores the result back into the location from which B was obtained. In addition, each instruction sets the four condition codes of the machine. However, the storage of the operation result and each of the four condition codes is conditional upon various flags within the instruction.

The operation code for the instruction consists of bits 24-27 (see in Figure 5-1). Operand A is specified by a register number (bits 0-3), an addressing mode (bits 4-5), and an immediate mode (bit 9). Operand B is specified by a register number (bits 10-13) and an addressing mode (bits 14-15). There are thus five addressing modes for operand A and four for operand B. The addressing modes are immediate (for operand A only), register direct, register indirect, register indirect with pre-decrement, and register indirect with post-increment. This is actually implemented by having two instruction formats, a two-address mode and an immediate mode, which are identical except for how the bits devoted to operand A are interpreted. See Figure 5-1. Each instruction contains a four bit "condition code mask" (bits 16-19) which specifies which of the four condition code registers are to be updated by the instruction. Finally, each instruction contains a four bit "store code" (bits 20-23) which specifies the conditions under which the output of the operation is to be stored into the effective address from which B was obtained.

TWO-ADDRESS MODE

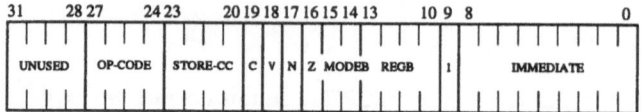

31	28 27	24 23	20 19 18 17 16 15 14 13	10 9 8	6 5 4 3	0

```
| UNUSED | OP-CODE | STORE-CC | C | V | N | Z | MODEB | REGB | 0 | UNUSED | MODEA | REGA |
```

IMMEDIATE DATUM MODE

31	28 27	24 23	20 19 18 17 16 15 14 13	10 9 8	0

```
| UNUSED | OP-CODE | STORE-CC | C | V | N | Z | MODEB | REGB | 1 | IMMEDIATE |
```

MODE	OPERAND	DESCRIPTION
00	Rn	Register Direct
01	(Rn)	Register Indirect
10	-(Rn)	Register Indirect Pre-decrement
11	(Rn)+	Register Indirect Post-increment

OP-CODE	MNEMONIC	OPERATION		STORE-CC	MNEMONIC	CONDITION
0000	MOVE	b <- a		0000	CC	~C
0001	INC	b <- a + 1		0001	CS	C
0010	ADDC	b <- a + b + c		0010	VC	~V
0011	ADD	b <- b + a		0011	VS	V
0100	NEG	b <- 0 - a		0100	PL	~N
0101	DEC	b <- a - 1		0101	MI	N
0110	SUBB	b <- b - a - c		0110	NE	~Z
0111	SUB	b <- b - a		0111	EQ	Z
1000	ROR	b <- c,a >> 1		1000	HI	~C & ~Z
1001	ASR	b <- a >> 1		1001	LS	C | Z
1010	LSR	b <- a >> 1		1010	GE	(N & V) | (~N & ~V)
1011	XOR	b <- b XOR a		1011	LT	(N & ~V) | (~N & V)
1100	OR	b <- b OR a		1100	GT	(N & V & ~Z) | (~N & ~V & ~Z)
1101	AND	b <- b AND a		1101	LE	Z | (N & ~V) | (~N & V)
1110	NOT	b <- NOT a		1110	T	True
1111	M15	b <- a		1111	F	False

Figure 5-1: The FM9001 Instruction Format

The FM9001 fetch-execute cycle is as follows:

1. The contents of register 15, the program counter, is treated as a memory address. The contents of that address is fetched and decoded into an opcode, store code, condition code mask, immediate mode, A and B addressing modes and A and B register numbers, as described above. Register 15 is then incremented by one.

2. Operand A is obtained as follows. This computation may also change the value of a register. If the immediate mode bit is on, then bits 0-8 of the instruction are sign extended to 32 bits and the result is operand A. Otherwise, let n be the register number specified for operand A and let v be the contents of that register. If the addressing mode for operand A is register direct, then v is used as operand A. If the mode is register indirect, the contents of memory location v is used as operand A. If the mode is register indirect with pre-decrement, the contents of memory location $v - 1$ is used as operand A and register n is set to $v - 1$. Otherwise, the mode must be register indirect with post-increment. In this case, the contents of memory location v is used as operand A and register n is set to $v + 1$.

3. Operand B is then obtained in a fashion exactly analogous to that for operand A, except that the immediate mode bit is ignored.

4. The operation indicated by the operation code of the instruction is performed by the ALU on operands A and B and the carry condition code, C. See Figure 5-1. The result is a 32-bit output and four output condition codes.

5. The four output condition codes are stored into their respective condition code registers provided the corresponding bit in the condition code mask is set.

6. If the store code of the instruction is satisfied by the output condition codes (see Figure 5-1), then the 32-bit output of the ALU is stored either into the register specified for operand B (if operand B was in register direct mode) or into the memory address from which B was fetched (if the B mode was anything but register direct).

5.2. Programming the FM9001

The mnemonic opcodes listed in Figure 5-1 do not include any jump or branch instructions. But because the program counter is a register, it can be conditionally set by any operation. For example, an unconditional jump can be coded as a MOVE instruction (opcode 1111) with store code T (1110) into register 15 (operand B addressing mode 00 and register number 1111) with operand A being the address to which we wish to jump. A conditional jump is one in which the store code is something besides T or F.

Let us take a concrete example. Suppose we wish to code a conditional jump to some fixed address x, where the condition tested is whether the carry flag is set.

One way to code "jump to x if carry set" is to put x into register 1, say, and use a conditional MOVE (store code CS, or "carry set" 0001) into register 15 from register 1. In binary this is

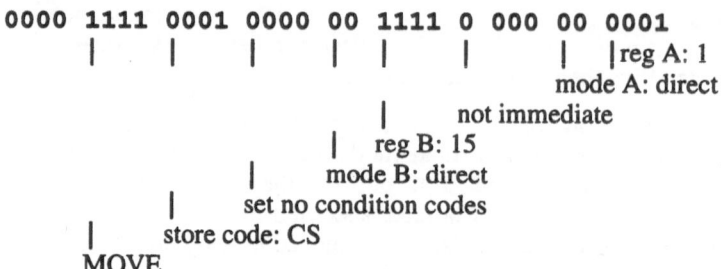

This instruction moves the contents of register 1, namely x, into the program counter register, provided the carry flag is set. Otherwise, it just increments the program counter by one. This style of coding a jump to a fixed address is inconvenient since it ties up an additional register.

It is more convenient to code such jumps so that the fixed address is written as part of the instruction stream, in particular, as the word immediately following the MOVE instruction. The coding problem therefore becomes "how can the next word in the instruction stream be treated as operand A?" This can be done by specifying operand A's addressing mode to be register indirect with post-increment and to specify register 15 (the program counter again) as the operand A register number.

Thus, another way to code "jump if carry set" to x is as a "double word" instruction, the first word being a conditional MOVE and the second word being x. The MOVE instruction, shown below, obtains operand A from the instruction stream by using register 15 in post-increment mode and relies upon the fact that the register will be incremented twice during the execution of the instruction.

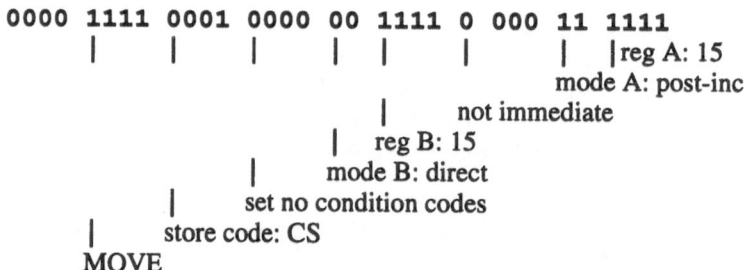

When execution of this MOVE instruction begins, register 15 points to it. But the first step in the fetch-execute cycle above increments register 15 after fetching this MOVE. Thus, when operand A is computed, register 15 points to the word after this MOVE, namely x. Because the addressing mode for operand A is register indirect with post-increment, x is used as operand A but register 15 is incremented (again) after fetching x. If the carry flag is set, x is thus moved into the program counter, as desired. If the carry flag is not set, the program counter (now incremented by two) points to the instruction after this double word instruction.

5.3. An FM9001 Assembly Language

In the course of implementing Piton on FM9001 we will develop an assembly language and assembler for FM9001. Even for the relatively short time-span it took to implement and prove Piton correct we could not write binary machine code accurately enough without mechanized aids. Thus, we will later discuss an assembler which allows us to write (**move-c** **()** **pc** (**pc** **+1**)) to assemble the machine code instruction discussed above. The assembler is formally defined and intimately involved in our proof of the correctness of the Piton implementation. Given the fact that it is possible to program the FM9001 in a symbolic assembly language, some readers might wonder why we developed Piton at all.

We therefore take a brief detour from our presentation of the FM9001 to contrast Piton with the FM9001. We have described Piton as an "assembly-level" language but we do not think of Piton as an assembly language for FM9001. The Piton machine is a higher level machine than the FM9001 machine. The former provides execute-only programs, subroutine call, local variables, stacks, and several data types; the latter provides 16 registers and 2^{32} words of memory. Generally speaking, the execution of a single Piton instruction requires the execution of several FM9001 instructions.

Recall that Piton was designed to let us move up from the machine code level to higher level languages and to do so in a way in which correctness proofs are convenient. Our assembly language for FM9001 does not solve these problems because the only abstractions it provides are symbolic opcodes, register names, and addressing modes. While those notions are helpful, they do not begin to approach the abstractions provided by Piton. FM9001 assembly code programs can necessarily overwrite themselves. Instructions are therefore necessarily data and hence are bit vectors. FM9001 assembly code programs must deal constantly with the fixed resources of the FM9001 machine. The verification of an FM9001 assembly language program, even one in which the registers and memory are used to implement stacks, must consider such low-level issues as the overwriting of program space, the allocations of registers, the protection of the stacks from "unstack-like" use, etc. Furthermore, these issues must be reconsidered every time a new assembly code program is verified.

But if we can implement Piton on the FM9001 and prove that we did it correctly, then these issues need not be considered again. Piton programs can be analyzed statically and (within the resource bounds declared by the user) can handle arbitrary numbers of recursive subroutine calls, local variables, and intermediate values. Furthermore, because of these features, higher level languages can be conveniently compiled into Piton. We now return to the main point of this chapter, which is the description of the FM9001.

5.4. Formalization and Verification

The FM9001 is formalized as the Nqthm function 'fm9001', which takes two arguments a machine state, *state*, and a number, *n*. The machine state is an n-tuple formalizing the registers, condition codes, and memory described above. 'Fm9001' returns the final state obtained by carrying out the fetch-execute cycle *n* times starting from *state*. The fetch-execute cycle of the FM9001 is formalized in the function 'fm9001-step', which takes a state and the binary representation of the register number of the program counter, and returns the next state. (The fact that the program counter is a parameter of 'fm9001-step' reflects the fact that in lower level descriptions of the processor some provisions are made for interrupts and alternative program counters.)

Thus, the formal definition of 'fm9001' is

DEFINITION:
fm9001 (*state*, *n*)

=

if $n \cong 0$ then *state*
else fm9001 (fm9001-step (*state*, nat-to-v (**15**, reg-size)),$n - 1$) endif

When we say "FM9001" in this book, we mean the function 'fm9001'.

As noted in the introduction, the FM9001 has been "verified." By that we mean that a much lower level description of this machine code interpreter has been written in the Nqthm logic and proved to be "equivalent" to 'fm9001', in a suitable sense. The lower level description is based on a formalization of the Netlist Description Language (NDL) of LSI Logic, Inc. Traditionally, hardware description languages allow engineers to describe electronic circuits in a hierarchical way from "modules" having "inputs" and "outputs." The most primitive modules represent gates implementing the Boolean operations and state-holding devices. With such a language one can describe the interconnection (outputs to inputs) of many such modules, thereby creating a new module. The entirety of such a description is called a "netlist."

In Brock and Hunt's formal hardware description language [7], netlists are represented by Nqthm list constants. Brock and Hunt define an interpreter which gives "simulation semantics" to these netlists: the outputs and next states of the primitive modules are given by Boolean operations on the inputs and current state; the interpreter determines the outputs and next state of successive modules as by simulation. It is thus possible to prove theorems about a given netlist (under this model). One such theorem is that a certain module implements a 32-bit wide binary adder in the sense that its outputs, when interpreted as a 33-bit binary number, is the sum of the two binary numbers represented by the 64 inputs.

Brock and Hunt developed a netlist for the FM9001 and proved that, under their semantics, the netlist implemented the 'fm9001' function described above. The formal netlist was then mechanically converted to LSI Logic's NDL and delivered to LSI Logic, Inc., which produced a CMOS gate-array from it using traditional CAD tools. This brief description of the hardware verification work of Brock and Hunt sweeps aside many complicated hardware design issues, such as gate delay, timing, fanout, power requirements, and pad assignments, all of which had to be considered by Brock and Hunt and some of which were dealt with formally in their model. The reader should see [7] for a more detailed description verification and fabrication of FM9001.

Because of the work done to verify and fabricate the FM9001, one might understand three different things by the noun "FM9001." It can mean the formally defined function 'fm9001' sketched above. It can mean the netlist or its interpretation under the Brock-Hunt simulation semantics. It can mean the silicon device manufactured by LSI Logic, Inc.

Generally in this book when we say "FM9001" we mean the 'fm9001' function. Our correctness theorem for Piton relates the Piton semantics to the 'fm9001' semantics of the binary image generated by our downloader. We characterize this as saying "Piton is correctly implemented on the FM9001," although it remains to be seen whether that is a fair characterization. Because of the theorem that Brock and Hunt proved, it is possible to understand our work as proving that "Piton is correctly implemented on the FM9001 netlist." That is, we can chain together the FM9001 correctness theorem with the Piton correctness theorem to derive a theorem relating Piton semantics to the computation performed by the Brock-Hunt netlist under their netlist simulation. The Piton downloader generates a suitable binary image for the gate-level view of the machine. This view of our work allows one to forget entirely about the FM9001 machine language.

But as noted in the introduction to this book, one cannot prove a theorem about a physical object. The existence of the device fabricated by LSI Logic, Inc., gives a wonderful sense of reality to this enterprise; the commitment to build the device kept everybody honest and forced attention onto certain practical problems. But it is simply a mistake to think that we *proved* that Piton is correctly implemented on that piece of silicon.

6

The Correctness of Piton on FM9001

The implementation of Piton on FM9001 is via what we call a "downloader" defined as the function 'load' in the logic. The function takes a Piton state, a list of natural numbers, called *boot-lst*, and a number, called *load-addr*. It generates an FM9001 state or "binary image."

The last two parameters allow the user to specify the initial contents of the low part of the FM9001 memory and the address at which the Piton-specific bits are to be placed. The motivation behind these two aspects of the downloader was to enable the testing of Piton images on the fabricated device. From a theoretical perspective it suffices to use the empty *boot-lst* and *load-addr* 0, resulting in the Piton-specific bits being loaded starting at memory address 0. But in our actual test jig for the fabricated device, it was necessary to have some test-specific code resident in the low part of memory and to have the Piton data segment located in a part of memory that was monitored by the test apparatus. The downloader was defined so that it could be made to generate suitable images. These aspects of the Piton implementation are still quite unsettled, though they are considered completely formally in our proof.

The correctness theorem is a formalization of the commutative diagram in Figure 6-1. The shaded boxes represent states. The upper pair are Piton states and the lower pair are FM9001 states. The horizontal arrows represent the state transformations wrought by stepping the corresponding machines. The downwardly directed vertical arrow is the downloader and the upwardly directed vertical arrow is a "display" function.

The basic idea is that one has some initial Piton state, p_0, and one wishes to obtain the state produced by running Piton forward n steps. However, depending on one's point of view, the abstract Piton machine does not "really" exist or is too expensive or inefficient to use. The correctness theorem tells us there is another way: map the Piton state "down" to an FM9001 state, run FM9001, and then map the resulting state back "up." However, the situation is much more subtle than suggested by the diagram. (Indeed, even the direction of the rightmost vertical arrow in the diagram might be questioned. We discuss that question at the end of this chapter. See page 93.)

The correctness theorem may be paraphrased as follows. Suppose p_0 is some

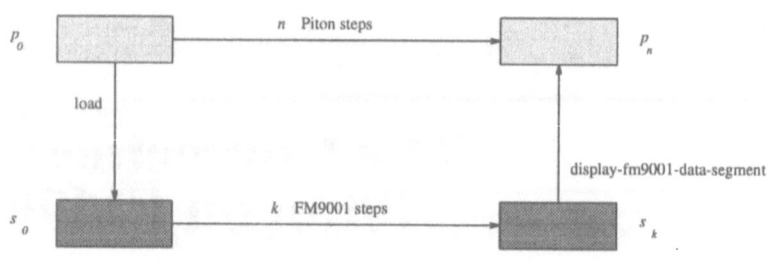

Figure 6-1: Piton on FM9001 Commutative Diagram

proper p-state that is loadable on FM9001 and has word size **32**. Let p_n be the p-state obtained by running the Piton machine n steps on p_0. Suppose p_n is not erroneous. Suppose furthermore that the "type specification" (see below) for the data segment of p_n is ts. Then it is possible to obtain the data segment of p_n via FM9001 as follows:

- *Down.* Let the initial FM9001 state, s_0, be load $(p_0, boot\text{-}lst, load\text{-}addr)$.

- *Across.* Obtain a final FM9001 state, s_k, by running FM9001 k steps on s_0, where k is a number obtained from p_0 and n by the constructive function 'fm9001-clock'.

- *Up.* Apply the constructive function 'display-fm9001-data-segment' to (a) the final FM9001 state, s_k, (b) the final type specification, ts, and (c) the link tables (computed by the constructive function 'link-tables' from p_0 and $load\text{-}addr$).

This is formalized as

THEOREM: FM9001 Piton is Correct
$$
\begin{aligned}
(\quad &\text{proper-p-statep}\,(p_0)\\
\wedge\quad &(load\text{-}addr \in \mathbf{N})\\
\wedge\quad &\text{p-loadablep}\,(p_0, load\text{-}addr)\\
\wedge\quad &(\text{p-word-size}\,(p_0) = \mathbf{32})\\
\wedge\quad &(p_n = \text{p}\,(p_0, n))\\
\wedge\quad &(\neg\,\text{errorp}\,(\text{p-psw}\,(p_n)))\\
\wedge\quad &(ts = \text{type-specification}\,(\text{p-data-segment}\,(p_n))))\\
\rightarrow\ (\quad &\text{p-data-segment}\,(p_n)\\
=\quad &\text{display-fm9001-data-segment}\,(\text{fm9001}\,(\text{load}\,(p_0, boot\text{-}lst, load\text{-}addr),\\
&\hspace{8em} \text{fm9001-clock}\,(p_0, n)),\\
&\hspace{4em} ts,\\
&\hspace{4em} \text{link-tables}\,(p_0, load\text{-}addr)))
\end{aligned}
$$

We believe this theorem is merely a formalization of what is usually meant by the informal remark that a programming language is implemented correctly on a given hardware base. Nevertheless (or perhaps, *because* it is an accurate formalization), as a correctness result the theorem is very subtle. What does this statement tell the user of FM9001 Piton? The rest of this chapter is devoted to a discussion of the meaning of this theorem. We discuss each of the hypotheses of the theorem and then what the conclusion tells us. We then apply the theorem to the **big-add** program of Chapter 4 to answer the question "Can we use FM9001 to add big numbers?" During that discussion we reconsider each of the hypotheses and the conclusion.

6.1. The Hypotheses of the Correctness Result

6.1.1. Proper P-States

In the first place, we are interested only in proper p-states. Those are the p-states in which all components are syntactically well-formed and compatible. This is a static, syntactic requirement, not a dynamic one. It includes such conditions as that every subroutine called be defined and that every data object (**addr** (*name* . *i*)) mentioned in the state be legal in the state (i.e., *name* is the name of a declared data area in the data segment and *i* is less than the length of that area). It does not include such dynamic conditions as that sufficient stack space exist for every subroutine call.

We could have chosen to specialize the correctness theorem still further and require that p_0 be an "initial" state. Without loss of computing power, we could define an initial state to be one in which the temporary stack is empty and we are about to execute the first instruction in the subroutine named **main**, (which has no local variables). We did not do this only because the proof requires a more general treatment of the mapping down from Piton to FM9001 and having paid for it we decided to provide it to the user.

6.1.2. Load-addr

The next hypothesis requires that *load-addr* be a natural number. This is no burden on the user.

6.1.3. Loadable

The third hypothesis requires that p_0 be "loadable" starting at address *load-addr* in the sense that *load-addr* plus the total size of the compiled programs, stacks, and data must not exceed the memory capacity of the FM9001. Thus, the FM9001 implementation of Piton is correct only for a subset of the abstract Piton state space. The abstract Piton machine has no inherent resource limitations—the user is free to specify any desired stack sizes. The concrete FM9001, on the other hand, is fundamentally limited to at most 2^{32} words of memory. Obviously, by specifying a huge stack size, the Piton programmer can generate an unloadable Piton state.

However, the situation is a little more subtle than that. The determination of whether a particular state is loadable is dependent not just on the resource limitations declared by the user but also on the amount of code generated by our compiler. This immediately contaminates the hypothesis of the correctness theorem with implementation details. Ideally one would like the correctness theorem for a programming language implementation to read something like "if the abstract state satisfies these conditions (all expressed in familiar abstract terms), then the computation can be carried out on the concrete machine as follows ..." But here we have a condition on the abstract state that, if one delves into its formal definition, involves such implementation details as how many words of machine code are generated for the **push-constant** instruction. We find this unaesthetic but unavoidable. How quickly the finite resources of the concrete machine are exhausted depends on how they are used by the implementation.

6.1.4. Word Size 32

Recall that the definition of Piton is parameterized by the word size. The word size is an explicit part of the Piton state and the abstract Piton machine is sensibly defined for all word sizes. However, the correctness theorem hypothesizes that the word size is 32. That is, the FM9001 implementation of Piton is correct only for word size 32.

This reflects the fact that the FM9001 is a 32-bit wide machine. Of course, our implementation could have allotted two FM9001 words to each Piton object and thus implemented word size 64. Indeed, the implementation could have been a nontrivial function of the word size and allotted as many FM9001 words as necessary to accommodate the user's declared word size, with the concomitant complications in the implementation of arithmetic and all other operations. We mention these possibilities only to alert the reader to the fact that the FM9001 implementation of Piton is *unnecessarily* restricted to a slice of the abstract state space. Nevertheless, such restrictions are standard practice in language implementations.

6.1.5. Non-Erroneous Final State

The next hypothesis is $p_n = p(p_0, n)$ which actually imposes no constraint but rather should be read as "let p_n be $p(p_0, n)$." The hypothesis after that one assumes that the final Piton state, p_n, is non-erroneous. That is, were this program run on the abstract Piton machine, the final psw would be either 'run or 'halt. But we are imagining that the program was run on FM9001, not the abstract Piton machine. Can we tell by looking at the FM9001 binary image whether the abstract machine would have caused an error? No. The only way this hypothesis can be relieved is for the user to know that the program runs without error. One way this knowledge can be gained is to (a) prove that the program runs without error on data satisfying its input conditions and (b) prove that the data used in this particular run satisfies those conditions. If the input conditions are expressed in an executable logic, then (b) can be done by computation. Alternatively, it is possible that the data is generated by a mechanism that can be proved always to generate satisfactory input.

Most modern compilers have a switch that allows the user to select whether the generated code is to be "efficient" or "safe." If "safe," the compiler generates runtime checks to insure that all the preconditions for every operation are met. When the compiler generates "efficient" code, no provision is made for those checks. If such code is executed on inappropriate input, no guarantees are offered. In particular, delivering an answer is not to be taken as an indication that the computation was error-free. The Piton compiler generates "efficient" code and has no switch causing it to generate "safe" code. Such a switch would be a nice improvement to the implementation but we felt that the first priority should be to generate efficient code since "safety" could be insured by proof.

6.1.6. Knowledge of the Final Type Specification

The next assumption is that *ts* is the type specification of the final Piton state. We have not discussed type specifications before. A *type specification* for a data segment is a structure isomorphic to the data segment except that where the data segment has a data object the type specification has just the type of the object. For example, the type specification for the data segment

```
((len (nat 5))
 (a    (nat 0) (nat 1) (nat 2) (nat 3) (nat 4))
 (x    (int -23)
       (nat 7)
       (bool t)
       (bitv (1 0 1 0 1 1 0 0))
       (addr (a . 3))
       (pc (setup . 25))
       (subr main)))
```

is

```
((len nat)
 (a    nat nat nat nat nat)
 (x    int
       nat
       bool
       bitv
       addr
       pc
       subr)).
```

In order to reconstruct the data segment of the final Piton state, the user must *know* the type specification of that state. "Whoa! Do you mean that as a programmer I have to know the type of every location in my final Piton data segment to read the answers from the FM9001 binary image?" Yes, you do. "The FM9001 binary image doesn't tell me what the types of the objects are?" No, it doesn't. "Then if I

don't have an abstract Piton machine, how can I know what the final types are?''
Proof.

This is not a new idea, but it looks a little startling when expressed formally.
When the assembly language programmer gets the answer

11111111110110100101001011110011

what does it mean? The answer might be -1,234,567 (if the bit vector represents an
integer in twos complement notation) or 4,293,732,729, or perhaps it means 10 of 32
items have some property. Interpreting the answer requires understanding the
program.

The idea of *ts* is startling because it is defined in terms of p_n, the supposedly
unknown final state. A naive view of the correctness theorem suggests the following
paradoxical reading: to determine the final data segment the programmer must
obtain the final data segment, get its type specification, and then inspect the binary
image. That is not the recommended approach!

How can one know the final type specification without obtaining the final data
segment? One can prove that one's interpretation of the final state is correct. For a
Piton program to be useful via the FM9001 implementation, one must not only prove
that it does not cause errors when called as expected but that it delivers a final data
segment with the expected type specification. This is not an onerous task. It is
usually part of the specification anyway. We illustrate this when we apply the
correctness result to **big-add** later in this chapter (page 86).

If Piton had a strongly typed syntax—so that it were impossible to change the
type of a variable or data location—then the final type specification would be the
same as the initial one and the theorem would no longer even suggest that p_n is
needed to recover the data segment of p_n. This suggests another approach using
FM9001 Piton: use only those Piton programs that do not change their type
specifications and prove that they have that property.

6.2. The Conclusion of the Correctness Result

The above discussion may be summarized as follows: The Piton programmer must
know that the program is proper, that the load address is a natural number, that the
static size of the FM9001 image is not excessive, that the word size is 32, and that
the final state is non-erroneous on this execution. The programmer must also know
the final type specification of the data segment. What does the conclusion then tell
the programmer?

6.2.1. The Final Data Segment

First of all, it does not permit reconstruction of the entire final state, only its data
segment. For example, we do not say how to recover from the final FM9001 state
the program segment of the final p-state. This is impossible (without some additional information) since, for example, the symbolic names of the Piton programs,

data areas, variables, and labels are all discarded by 'load'. Furthermore, it is pointless to reconstruct the program segment: it never changes anyway. More questionable, perhaps, is that we do not show how to recover the final control or temporary stacks.

We felt it was sufficient to handle the data segment alone. If one has a program whose final answer is left on the stack, one could add the additional Piton code to move that answer into the data area. That is what we did in the **main** program (page 48): after calling **big-add** we pop the ''carry out'' flag off the temporary stack and store it into the global variable **c**. Thus, the implementation correctness theorem enables us to get both of **big-add**'s answers (the final array and the flag) out of the FM9001 binary image.

6.2.2. The FM9001 Route

If all of the hypotheses are true for a particular run, this theorem explains how to get the final Piton data segment given just the FM9001, the initial state, p_0, n, and an arbitrary *boot-lst* and numeric *load-addr*. In particular, it is

display-fm9001-data-segment (fm9001 (load $(p_0,$ *boot-lst*, *load-addr*),

fm9001-clock $(p_0, n))$,

ts,
link-tables $(p_0,$ *load-addr*))

The functions, 'load', 'link-tables', and 'display-fm9001-data-segment' are all defined in the formal treatment of the correctness result. 'Fm9001-clock' is discussed below and on page 299.

Informally, 'load' constructs an FM9001 binary binary image from a Piton state by compiling, assembling and linking the Piton programs and data areas. 'Fm9001-clock' determines how many clock ticks it takes FM9001 to carry out the computation performed by Piton in n ticks—a determination made by carrying out the computation with the abstract Piton machine and counting how many FM9001 instructions are used in our implementation. This may strike some readers as problematic and we will return to it below. 'Link-tables' is part of the 'load' function. It takes a Piton state and the load address and computes several tables that tell the linker where each program, label, stack, and data area is to be located in absolute memory space.

'Display-fm9001-data-segment' reconstructs the Piton data segment by scanning the type specification *ts* to determine the names of the data areas, using the link tables to determine the absolute location at which each data area was allocated, scanning the memory of the final state from those locations to recover bit vectors, and finally using the type specification again to convert the individual bit vectors back into Piton data objects of the appropriate type. The formalization of this concept is embodied in the function 'display-fm9001-data-segment' which is defined on page 301.

We suggest that this is exactly what the assembly language programmer (or debugger) does when inspecting a binary dump.

6.3. The Termination of FM9001

Our handling of the termination of Piton programs on FM9001 leaves something to be desired. As things stand now, the correctness theorem tells us that the FM9001 state on the kth clock tick maps up as described, where k is fm9001-clock (p_0, n). But the theorem says nothing about what FM9001 does on the $k+1^{st}$ clock tick. Indeed, the answers computed by the Piton code may well be destroyed by subsequent FM9001 computation. It therefore seems crucial that we be able to determine k and then look at the FM9001 state at precisely the right step.

But the only way to determine k is to use 'fm9001-clock', which runs the computation on the abstract Piton machine. If we assume that it is impractical to run the abstract Piton machine to do the original computation, it is surely impractical to run it to count steps for FM9001. However, we do not intend 'fm9001-clock' to be executed. Indeed, we prefer to read the correctness theorem as a constructive way of saying "there exists a number k such that if FM9001 is run k ticks a suitable state results." 'Fm9001-clock' is a "witness function" that exhibits a suitable k. In a suitable logic, such as Kaufmann's extension to Nqthm-1992 [21] to support first-order quantifiers, 'fm9001-clock' would not appear in the statement of the correctness theorem. Instead, it would appear as an artifact of the proof we checked. While k is constructively given in the proof via the total recursive function 'fm9001-clock', no practical computational means is given for determining its value in an actual application. Because we regard 'fm9001-clock' merely as a witness function we do not exhibit a definition of 'fm9001-clock' in this book, though we discuss it again on page 299.

One must therefore ask whether the correctness result offers a path to practical Piton computation on the FM9001? The answer is yes, even though, for many practical purposes 'fm9001-clock' cannot be evaluated. The Piton user could write the top-level Piton program, e.g., our **main**, to set some global variable, say **terminated**, to t and enter an infinite loop when the computation has completed. Then the user could prove that whenever **terminated** is t the answer then visible is correct. It would then be convenient to execute such a Piton image on the FM9001 because once **terminated** becomes true, the data no longer changes and is known to be correct.

The main reason that the issue of the clock has not been more conveniently addressed in our correctness theorem is we do not really know what we will require of FM9001 Piton. We expect that a more sophisticated treatment of termination to evolve.

6.4. Applying the Correctness Result to Big-Add

Recall that in Chapter 4 we defined a small system of two Piton programs for doing big number addition. Suppose we are now interested in running that system on the FM9001 implementation of Piton. What does the correctness result tell us?

The system in question is constructed by the function

DEFINITION:
system-initial-state (a, b)
=
p-state $('$ **(pc (main . 0))**,
 $\quad\quad\quad '$ **((nil (pc (main . 0)))))**,
 $\quad\quad\quad$ **nil**,
 $\quad\quad\quad$ list (main-program, big-add-program),
 $\quad\quad\quad$ list (cons $('$ **bna**, a),
 $\quad\quad\quad\quad\quad$ cons $('$ **bnb**, b),
 $\quad\quad\quad\quad\quad$ cons $('$ **n**, list (tag $('$ **nat**, length (a)))),
 $\quad\quad\quad\quad\quad$ cons $('$ **c**, list (tag $('$ **nat**, 0)))),
 $\quad\quad\quad$ **10**,
 $\quad\quad\quad$ **8**,
 $\quad\quad\quad$ **32**,
 $\quad\quad\quad$ '**run**)

We defined the acceptable values of a and b with

DEFINITION:
system-initial-state-okp (a, b)
=
 $\quad\quad$ bignp $(a, 2^{32})$
 $\wedge\quad$ bignp $(b, 2^{32})$
 $\wedge\quad$ (length (a) = length (b))
 $\wedge\quad$ ($\neg$ (length $(a) \cong$ **0**))
 $\wedge\quad$ (length $(a) < 2^{32}$)

On page 61 we proved the following theorem about this Piton system:

THEOREM:
 $\quad$ system-initial-state-okp (a, b)
 $\rightarrow$ ($\quad$ p (system-initial-state (a, b), system-initial-state-clock (a, b))
 $\quad$ = p-state $('$ **(pc (main . 5))**,
 $\quad\quad\quad\quad '$ **((nil (pc (main . 0)))))**,
 $\quad\quad\quad\quad$ **nil**,
 $\quad\quad\quad\quad$ list (main-program, big-add-program),
 $\quad\quad\quad\quad$ list (cons $('$ **bna**, big-add-array $(a, b,$ **f**, 2^{32})),
 $\quad\quad\quad\quad\quad\quad$ cons $('$ **bnb**, b),
 $\quad\quad\quad\quad\quad\quad$ cons $('$ **n**, list (tag $('$ **nat**, length (a)))),
 $\quad\quad\quad\quad\quad\quad$ cons $('$ **c**, list (big-add-carry-out $(a, b,$ **f**, 2^{32})))),
 $\quad\quad\quad\quad$ **10**,
 $\quad\quad\quad\quad$ **8**,
 $\quad\quad\quad\quad$ **32**,
 $\quad\quad\quad\quad$ '**halt**))

Suppose we have in mind some particular big numbers a and b such that
system-initial-state-okp (a, b) holds. We know how to add them together with the
abstract Piton machine: let p_0 be system-initial-state (a, b) and let p_n be the result of
running the Piton machine system-initial-state-clock (a, b) steps starting from p_0.
Then by the correctness of the **big-add** system (above) we know that the data
segment of p_n contains the correct big number sum. Can we get that same answer
via FM9001?

It is not at all obvious that FM9001 Piton can do the job for us, given the complexity of our theorem stating the correctness of FM9001 Piton. We must know six things. First, we must know that p_0 is a proper p-state. Second, we must select some natural number for the *load-addr*. Third, we must know that p_0 is loadable. Fourth, we must know that the word size of p_0 is **32**. Fifth, we must know that the final state, p_n, is non-erroneous. Sixth, we must know the type specification of the data segment of p_n. We deal with each of these issues in turn. Recall that p_0 is system-initial-state (a, b) and we know system-initial-state-okp (a, b).

6.4.1. Proper P-States

We can prove that p_0 is proper from the assumption that a and b satisfy 'system-initial-state-okp'. In particular,

THEOREM:

system-initial-state-okp $(a, b) \rightarrow$ proper-p-statep (system-initial-state (a, b))

is straightforward to prove. Recall that 'proper-p-statep' is concerned with such issues as whether the programs in the program segment are syntactically well-formed and mention no global data other than that declared in the data segment and that all of the Piton objects involved in the state are legal. But the program segment in this case is a constant, namely the list containing the **main** program and **big-add**, and the data segment is "almost" a constant—the names of the areas are explicitly given and their contents are derived from a and b. So it is easy to confirm that our programs are proper and only refer to declared data. It is interesting that this is the first place in the **big-add** exercise where we are concerned with the question of whether our programs are syntactically well-formed. In addition to the syntactic checks discussed above, 'proper-p-statep' checks that all Piton objects in the initial state are legal. The proof that this is true for system-initial-state (a, b) appeals to the 'system-initial-state-okp' hypothesis, which guarantees that a and b are big numbers, and hence the **bna** and **bnb** data areas contain legal Piton natural numbers.

6.4.2. Load-Addr

We are free to select any natural number for *load-addr*, though if we select too big a number the next hypothesis, that the state is loadable at that address, will fail. We will choose **0**.

6.4.3. Loadable

The next question is whether p_0 is loadable (starting at address *load-addr*, which is here **0**). That depends on how big a and b are. If each is of length 2^{100}, p_0 is certainly not loadable since its data segment contains two areas each of which require 2^{100} words. On the other hand, for "small" a and b the system initial state is loadable. We have proved mechanically

THEOREM:

(system-initial-state-okp (a, b) $\wedge$ (length (a) < **1000**))
$\rightarrow$ p-loadablep (system-initial-state (a, b), **0**)

which guarantees that if a has fewer than a thousand digits then p_0 is loadable. We could have proved a much higher bound on the size of a and b but we felt that this bound makes the point that we can do significantly large additions with this system. Recall that each "digit" here is natural between **0** and 2^{32}.

Recall that 'p-loadablep' is implementation dependent. Thus, the proof of the above theorem actually involves reasoning about the data representation and the compiler. We sketch the proof here simply to reassure the reader that this "implementation dependent" reasoning does not require undue familiarity with the FM9001 implementation of Piton. The program segment created by 'system-initial-state' is constant and so its FM9001 size is constant and may be determined by computation to be **97**. (That is, if you call the compiler on **main** and **big-add** and add the lengths of the resulting programs you get **97**) The space allocated to the Piton stacks in this example is also constant and is seen to be **23**. The data segment constructed by 'system-initial-state' has four areas, one each for a and b (each containing length (a) words) and a word each for n and c. Thus the total size of the loaded system is **97** + **23** + **2** + (**2** $\times n$), where n is length (a). The theorem has thus been reduced to showing that if n<1000 then (**122** + (**2** $\times n$)) < 2^{32}. This is trivial.

6.4.4. Word Size 32

The fourth issue is the word size of p_0. It is easy to prove

THEOREM:
p-word-size (system-initial-state (a, b)) = **32**

since 'system-initial-state' explicitly supplies a word size of **32**.

6.4.5. Non-Erroneous Final State

The fifth issue is establishing that the final state, p_n, is non-erroneous. Recall that p_n is
p (system-initial-state (a, b), system-initial-state-clock (a, b))

and recall the correctness of the **big-add** system. From that result we know that the final psw of this state is in fact **'halt** and so the state is non-erroneous. More generally we can prove mechanically

THEOREM:

system-initial-state-okp (a, b)
$\rightarrow$ ($\neg$ errorp (p-psw (p (system-initial-state (a, b),
      system-initial-state-clock (a, b))))))

Thus, the fact that the FM9001 implementation of Piton is correct only for non-erroneous computations is not a problem because we are using it to run verified code and we proved that our code does not cause errors when run on input satisfying our input conditions.

6.4.6. Knowledge of the Final Type Specification

The sixth issue is knowledge of the type specification of the final data segment. This was perhaps the most problematic aspect of our implementation correctness theorem. Recall that the problem is that to recover the final data segment from the final FM9001 binary image it is necessary to *know* the types of the objects in the final data segment. But it is clearly impractical to obtain that knowledge by computing the final data segment with the abstract Piton semantics since there would then be no need to compute with FM9001. We argued that the final type specification could be obtained via proof.

The following mechanically proved theorem illustrates this argument.

THEOREM:
    system-initial-state-okp (a, b)
$\rightarrow$ (type-specification (p-data-segment (p (system-initial-state (a, b),
                                       system-initial-state-clock (a, b)))))
  = list (cons (**'bna**, listn (length (a), **'nat**)),
        cons (**'bnb**, listn (length (a), **'nat**)),
        cons (**'n**, list (**'nat**)),
        cons (**'c**, list (**'bool**))))

where

DEFINITION:
listn (n, x)
=
if $n \cong 0$ **then nil**
else cons $(x$, listn $(n-1, x))$ **endif**

'Listn' constructs a list of n occurrences of x Thus, for example, listn (**4**, **'nat**) is **'(nat nat nat nat)**.

The theorem above tells us that for a and b satisfying 'system-initial-state-okp' the type specification of the data segment of the final abstract p-state is

list (cons (**'bna**, listn (length (a), **'nat**)),
    cons (**'bnb**, listn (length (a), **'nat**)),
    cons (**'n**, list (**'nat**)),
    cons (**'c**, list (**'bool**)))

which means the final value of the **bna** array is a list of natural numbers. The final value of the **bnb** array is also a list of natural numbers. The final value of :a is a natural. The final value of **c** is a Boolean. Note that **c** changes type in the computation. Its initial value was a natural. This theorem is easy to prove from the correctness of the **big-add** system (i.e., from the specification of **big-add**—there is no need to inspect the code of the subroutine). After all, according to our correctness result, the final value of **bna** is a big number and so is a list of naturals, the final value of **c** is the carry out, which is a Boolean, and **bnb** and **n** are shown not to have changed. Since the system is correct, we know the final type specification even without knowing the final data segment.

6.4.7. Using FM9001 to Add Big Numbers

Given all of the foregoing we know that if a and b satisfy 'system-initial-state-okp' then FM9001 can be used to add them together. Another way to view the situation is that we can combine the correctness of the **big-add** system with the correctness of the FM9001 implementation of Piton and eliminate entirely the semantics of Piton.

The theorem is

THEOREM:
$$(\text{system-initial-state-okp}\,(a,\,b) \wedge (\text{length}\,(a) < \mathbf{1000}))$$
$$\rightarrow (\quad \text{display-fm9001-data-segment}\,($$
$$\text{fm9001}\,($$
$$\text{load}\,(\text{system-initial-state}\,(a,\,b),\,\textbf{nil},\,\mathbf{0}),$$
$$\text{fm9001-clock}\,(\text{system-initial-state}\,(a,\,b),$$
$$\text{system-initial-state-clock}\,(a,\,b))),$$
$$\text{list}\,(\text{cons}\,(\textbf{'bna},\,\text{listn}\,(\text{length}\,(a),\,\textbf{'nat})),$$
$$\text{cons}\,(\textbf{'bnb},\,\text{listn}\,(\text{length}\,(a),\,\textbf{'nat})),$$
$$\text{cons}\,(\textbf{'n},\,\text{list}\,(\textbf{'nat})),$$
$$\text{cons}\,(\textbf{'c},\,\text{list}\,(\textbf{'bool}))),$$
$$\text{link-tables}\,(\text{system-initial-state}\,(a,\,b),\,\mathbf{0}))$$
$$= \quad \text{list}\,(\text{cons}\,(\textbf{'bna},\,\text{big-add-array}\,(a,\,b,\,\textbf{f},\,2^{32})),$$
$$\text{cons}\,(\textbf{'bnb},\,b),$$
$$\text{cons}\,(\textbf{'n},\,\text{list}\,(\text{tag}\,(\textbf{'nat},\,\text{length}\,(a)))),$$
$$\text{cons}\,(\textbf{'c},\,\text{list}\,(\text{big-add-carry-out}\,(a,\,b,\,\textbf{f},\,2^{32}))))))$$

Observe that the theorem does not mention 'p', proper p-states, erroneous computations or type specifications. It says that if you have acceptable big numbers a and b then you can add them together by creating an FM9001 state, running FM9001, and then applying 'display-fm9001-data-segment' to the final state.

As stated above the theorem is slightly unsatisfying because the second and third arguments to 'display-fm9001-data-segment' still involve a and b. Why is this unsatisfying? Because one has to look carefully at the way a and b are used there to convince oneself that the big-number addition work is being done by the FM9001 and not by 'display-fm9001-data-segment' or its use. For example, how do we know that 'display-fm9001-data-segment' is not just ignoring the final FM9001 state, recovering the initial a and b somehow from the type specification and link tables and using our abstract 'big-plus' to compute the answer? To dismiss this possibility one must delve into a number of complicated functions, including 'display-fm9001-data-segment' and 'link-tables'.

But we "know" that the FM9001 *is* doing the work and we can make the theorem above more clearly express that. In fact, the expressions in the second and third argument positions of 'display-fm9001-data-segment' are functions only of the length of a. That is, given the length of a we can construct the type specification and the link tables without any further knowledge of the particular a or b to be added. We can thus define the function 'display-answers' so that it takes only the final FM9001 state and the length of the big number system and returns the list consisting of the sum and carry out. 'Display-answers' has built into it the types and absolute locations of the answers in the final binary image. We can then repackage the above into

THEOREM:
   (system-initial-state-okp $(a, b) \wedge$ (length $(a) <$ **1000**))
$\rightarrow$ (display-answers (
            fm9001 (load (system-initial-state (a, b), **nil**, **0**),
                  fm9001-clock (system-initial-state (a, b),
                              system-initial-state-clock (a, b)))),
         length (a))
   =   list (big-add-array $(a, b, \mathbf{f}, 2^{32})$, big-add-carry-out $(a, b, \mathbf{f}, 2^{32})$)))

The above theorems demonstrating that FM9001 can do big number arithmetic would be hard to prove in the absence of our results on Piton, of course. Indeed, the beauty of the current arrangement of theorems is that the difficult, problem specific reasoning that establishes our programs correct is done with respect to the elegant abstract semantics of Piton. But our programs can be efficiently executed by the concrete FM9001.

It is often difficult to ascertain whether a given formal sentence adequately captures the informal notions offered in explanation. Does our correctness result for the FM9001 implementation of Piton actually capture the alleged idea? Without a formal "theory of implementations" that question cannot be answered. However, we can "stack" the correctness of a given Piton program on top of the correctness of FM9001 Piton to get a theorem that eliminates the intermediate level, i.e., Piton. We offer this as evidence that our correctness theorems are adequate.

6.4.8. Concrete Data

We now return to the question that started our analysis of **big-add** versus the correctness result. Imagine that we have two particular big numbers we wish to add, say '((**nat 246838082**) (**nat 3116233281**) (**nat 42632655**) (**nat 0**)) and '((**nat 3579363592**) (**nat 3979696680**) (**nat 7693250**) (**nat 0**)). Can we use FM9001 to add them? We now know the answer is "yes" provided the two numbers satisfy 'system-initial-state-okp' and the numbers have fewer than one thousand digits. The theorem

THEOREM:
```
    system-initial-state-okp('((nat  246838082)
                               (nat  3116233281)
                               (nat  42632655)
                               (nat  0)),
                             '((nat  3579363592)
                               (nat  3979696680)
                               (nat  7693250)
                               (nat  0)))
∧  (    length('((nat  246838082)
                 (nat  3116233281)
                 (nat  42632655)
                 (nat  0)))
    <  1000)
```

is trivial to prove by computation. Thus, if we run FM9001 on the corresponding initial state and extract the answers they will be correct. The point is that despite all our analysis and proofs, we know that the final answer is correct only if we know we are using the program in accordance with its specifications.

One last point deserves to be made here. Since we did not design the big number addition system to signal its own termination we must know, for each concrete input data set, how long to let the machine run. For big numbers a and b, the FM9001 must run for

fm9001-clock (system-initial-state (a, b), system-initial-state-clock (a, b))

instructions. For the concrete data above, this expression is equal to **190**.

In fact, it turns out that in our implementation, any pair of big numbers of length **4** are added in **190** FM9001 instructions. That is, the number of FM9001 instructions used is a function only of the length of the data. But some Piton instructions compile into blocks of code involving branches and thus this nice state of affairs— that 'fm9001-clock' is data independent—does not generally hold.

6.5. Upwards versus Downwards

We conclude this chapter on the meaning and use of the Piton correctness theorem by returning to the commutative diagram offered in explanation. As noted, the correctness theorem is a formalization of the commutative diagram in Figure 6-2.

A frequently asked question concerns the upward direction of rightmost vertical arrow the diagram. In order to state the theorem this way we have to invent a new function, 'display-fm9001-data-segment', which represents that arrow. A more natural diagram, perhaps, is one with two downwardly directed arrows, as shown in Figure 6-3. Now both arrows are formally represented by 'load' and the statement of the problem is simpler.

But while the theorem corresponding to Figure 6-3 is easier to state, it does not obviously establish that Piton is implemented on FM9001. For example, a definition of 'load' with the property pictured is one that maps all Piton states into a one instruction infinite loop, i.e., a state in which the FM9001 is halted. If we let s_0 be

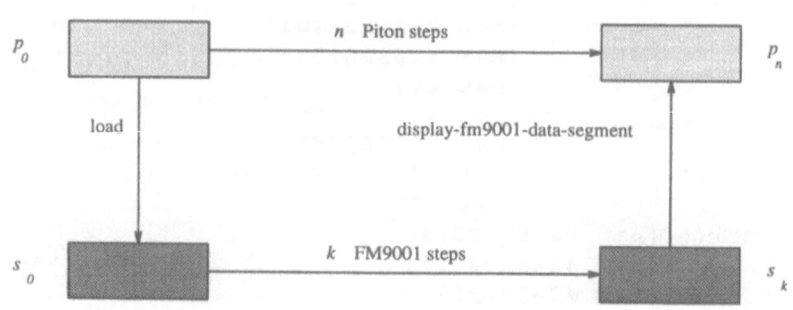

Figure 6-2: Piton on FM9001 Commutative Diagram

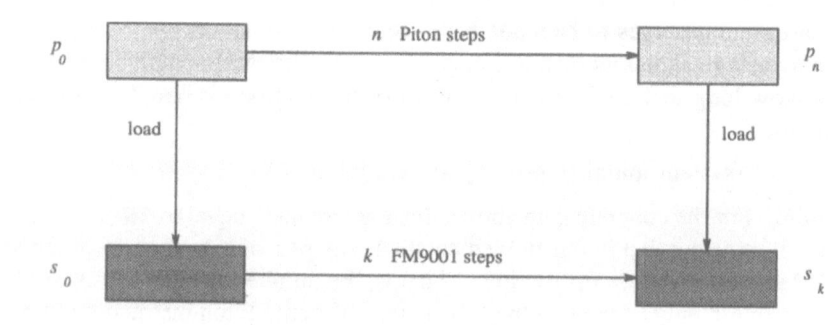

Figure 6-3: An Alternative Commutative Diagram

such a state, then running FM9001 forward k steps from s_0 produces s_k which is s_0. Thus, fm9001 (load (p_0), k) and load (p $(p_0$, $n)$) are equal, as required by Figure 6-3. See Figure 6-4.

The theorem corresponding to our diagram does not suffer this problem. If you are given a 'load' and a 'display-fm9001-data-segment' having the relationship described in our diagram, and you have FM9001, you can do Piton computations. Put another way, our diagram shows that there is a path from p_0 to p_n that does not involve the formal definition of Piton. The simpler diagram only offers two paths to a possibly uninteresting lower level state.

It turns out that to prove the correctness theorem we first prove a lemma very much like Figure 6-3 and then prove that 'load' is "inverted" by 'display-fm9001-data-segment'. But what has concerned us in this chapter is the development of a convincing *statement* of the correctness property.

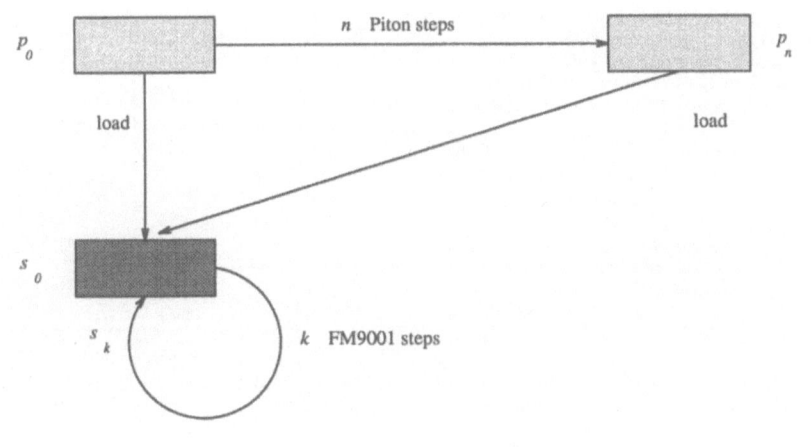

Figure 6-4: A Satisfactory 'Load'?

7

The Implementation of Piton on FM9001

To implement Piton on FM9001 we define 'load'. 'Load' takes a p-state, a list of natural numbers, *boot-lst*, and a load address, *load-addr*, as its input and produces an FM9001 state as its output. 'Load' carries out four successive transformations starting with the initial p-state and ending with an FM9001 state. Each such transformation is described by a function and 'load' is the composition of those four functions. The four functions have names that suggest they are transforms from one state space to another. For example, the function 'p⇒r' transforms p-states to what we call "r-states." We explain these intermediate forms later.

DEFINITION:
load (*p*, *boot-lst*, *load-addr*)
=
m⇒fm9001 (i⇒m (r⇒i (p⇒r (*p*)), *boot-lst*, *load-addr*))

The four phases of 'load' are summarized below.

- The first phase (implemented by the innermost function, 'p⇒r') is the *resource representation* phase. In this step we are concerned with using the resources of the FM9001 to represent the stacks and other resources of the Piton machine. The principal output of this phase is a symbolic description of the contents of the FM9001 registers and that portion of the memory containing the stacks. This symbolic description will be subjected to "linking" (below). 'P⇒r' is defined formally on page 294.

- In the second phase ('r⇒i', formally defined on page 295), we translate Piton instructions into FM9001 instructions. We call this *compiling* the Piton programs. In our implementation, the compiler does not produce binary machine code; instead it produces what we call "i-code," a typed, relocatable assembly code for FM9001 containing symbolic program names, labels, and abstract Piton objects. The principal output of this phase is a new program segment in which each program name is associated with its i-code.

- In the third phase ('i⇒m', formally defined on page 271), we replace all of the symbolic objects—in the stacks, in the assembled programs, and in the data segment—by bit vectors. This is called *link-assembling*. The i-code instructions are *assembled* into their machine code counterparts. Piton data objects are *linked* to (i.e., replaced by) the corresponding bit vectors. We are justified in calling this process "linking" because our compiled i-code uses Piton data objects (e.g., program addresses) for all references. Thus, the link phase is responsible for tying the system of programs and data together with absolute addresses.

- Finally, in the fourth phase ('m⇒fm9001', formally defined on page 293), we convert our representation of the register file and memory to that used by the FM9001. The FM9001 uses binary trees to represent its register file and memory. For example, instead of representing the 16 registers as a list of 16 bit vectors, it represents them as a full balanced binary tree of depth 4, called a "ram tree."

The organization of this chapter is as follows. We first present an example p-state and the corresponding FM9001 state. This is offered primarily as proof that our implementation ultimately produces a "binary image" as it is traditionally understood, despite the fact that the compiler and link-assembler are written in a computational logic. Next we sketch the implementation, primarily to establish terminology. Then we describe each of the four phases of the implementation: resource representation, compiling, link-assembling, and image construction.

7.1. An Example

Recall the program **big-add** and the initial state illustrated in Figure 4-2 on page 50. The result of downloading this state is the FM9001 state described in Figure 7-1. There is absolutely no expectation on our part that this binary dump is understandable. We only expect it to impress upon the reader that 'load' does indeed produce a binary image, despite our early fascination with abstractions such as "resource representation."

Registers:
```
 0:00000000000000000000000000000000 00000000000000000000000001110011
 2:00000000000000000000000001110011 00000000000000000000000001111110
 4:00000000000000000000000000000000 00000000000000000000000000000000
 6:00000000000000000000000000000000 00000000000000000000000000000000
 8:00000000000000000000000000000000 00000000000000000000000000000000
10:00000000000000000000000000000000 00000000000000000000000000000000
12:00000000000000000000000000000000 00000000000000000000000000000000
14:00000000000000000000000000000000 00000000000000000000000000001100
```
Flags (Z N V C):
```
 0 0 0 0
```
Memory:
```
 0:00001110101101100111001101000010 10111001101111011111001001000001
 2:00000010100010101000010111001111 00000000000000000000000000000000
 4:11010101010110001100000100001000 11101101001101010101101000101000
 6:00000000011101010110001111000010 00000000000000000000000000000000
 8:00000000000000000000000000000000 00000000000000000000000000000000
10:00001111111000010001000000000001 00001111111000000000010000000010
12:00001111111000010001100001111111 00000000000000000000000000000000
```

```
 14:000011111110000010001100001111110000000000000000000000000000000100
 16:000011111110000000010000001111110000000000000000000000000000001000
 18:000011111110000010001100000101000001111111100001000100000111111
 20:000000000000000000000000000101110000111111110000000111100000011111
 22:000000000000000000000000000011111000011111111000000001000000111111
 24:000000000000000000000000000001001000011111110000001010000001100111
 26:000011111110000000011110000111110000000000000000000000000011100
 28:000011111110000000001000000000010000111111110000000010000011001
 30:000011111110000000011110000110010000111111110000010001000000001
 32:000011111110000000001000000010000011111110000010001000001100011
 34:000011111110000010001000001100110000111111110000010001000001100011
 36:000011111110000010001100001111110000000000000000000000000000000
 38:000011111110000000010000001111110000000000000000000000000000000
 40:000000111110000000010000000000100001111111000001000110000100100
 42:000011111110000000010000001100110001111111000001000110000010100
 44:000011111110000000010000001111110000000000000000000000000000001
 46:000000111110000000010000000000100001111111000001000110000010100
 48:000011111110000000010000001100110001111111000001000110000010100
 50:000011111110000000010000001100110001111111000000001010000110011
 52:000010011101010000100110000010010000010111010000001000000000101
 54:000011110001000001001100001111110000000000000000000000000000001
 56:000011111110000010001100000001000001111111100000001000000111111
 58:000000000000000000000000000000000000001111100000001000000000010
 60:000011111110000010001100000101000001111111000000001000000110011
 62:000011111110000010100000001100110001111111000000001000000111111
 64:000000000000000000000000000000100000011111100000001000000000010
 66:000011111110000010001100000101000000010111000001001100000010011
 68:000011111110000000010000001111110000000000000000000000000000010
 70:000000111110000000010000000000100001111111000000101000000010011
 72:000011111110000100010100001100110001111111000000001000000111111
 74:000000000000000000000000001100110000111101110000011110000000100
 76:000011111110000000010000001111110000000000000000000000000000000
 78:000000111110000000010000000000100001111111000001000110000010100
 80:000011111110000010001100001111110000000000000000000000000000001
 82:000011111110000000010000001100110000001111100000100110000000100
 84:000011111110000000010000001111110000000000000000000000000000001
 86:000000111110000000010000000000100001111111000001010000000110011
 88:000011111110000000010000001111110000000000000000000000000000000
 90:000000111110000000010000000000100001111111100001000110000010100
 92:000011111110000010001100001111110000000000000000000000000000001
 94:000011111110000000010000001100110000001111100000100110000000100
 96:000011111110000000010000001111110000000000000000000000000000000
 98:000000111110000000010000000000100001111111100000101000000010011
100:000011111110000000011100000111110000000000000000000000000101010
102:000011111110000000011100000111110000000000000000000000001101000
104:000011111110000000001000000000010001111111100000000010000110010
106:000011111110000000011100001100100000000000000000000000000000000
108:000000000000000000000000000000000000000000000000000000000000000
110:000000000000000000000000000000000000000000000000000000000000000
112:000000000000000000000000000000000000000000000000000000000000000
114:000000000000000000000000000000000000000000000000000000001110011
116:000000000000000000000000000001100000000000000000000000000000000
118:000000000000000000000000000000000000000000000000000000000000000
120:000000000000000000000000000000000000000000000000000000000000000
122:000000000000000000000000000000000000000000000000000000000000000
124:000000000000000000000000000000000000000000000000000000000000000
126:000000000000000000000000000000000000000000000000000000001101011
128:000000000000000000000000001110110000000000000000000000001111110
```

Figure 7-1: The FM9001 Binary Image for Big Number Addition

The binary image in Figure 7-1 was generated by load(p_0, **nil**, **0**), where p_0 is the initial p-state for **big-add** as shown on page 50. The even numbered registers and memory locations are in the left-hand column and the odd numbered ones are in the right-hand column. We display the registers and memory linearly rather than in the form of ram trees, as required in an FM9001 state. Another technical discrepancy is that a bit vector in the FM9001 model is a list of Booleans and when such a list is interpreted as a natural number the least significant bit is that in the 'car' and thus appears on the left. Thus, in the FM9001 model, list (**t, f, f, f**) is the bit vector of length 4 representing the natural number **1**. But in Figure 7-1 we display bit vectors as sequences of **0**s (for **f**s) and **1**s (for **t**s) and we place the least significant bit on the right following conventional binary notation. Thus, under the conventions used in Figure 7-1, **0001** stands for list (**t, f, f, f**).[7]

7.2. A Sketch of the FM9001 Implementation

Our implementation of Piton uses six of the sixteen registers of the FM9001 and all four of the condition code flags. We partition the memory of FM9001 into four parts, called, respectively, the boot code, the user data segment, the program segment, and the system data segment.

7.2.1. The Registers

We give symbolic names to six of the FM9001's registers.

- Register 1 is called the *cfp* (control stack frame pointer) register. It indicates the extent of the topmost frame on the control stack.

- Register 2 is called the *csp* (control stack pointer) register. It addresses the top of the control stack.

- Register 3 is called the *tsp* (temporary stack pointer) register. It addresses the top of the temporary stack.

- Registers 4 and 5 are called the x and y registers, respectively. They are used as *temporaries* in the machine code implementing Piton instructions: no assumptions are made about the values of these registers when (the machine code for) a Piton instruction begins execution and no promises are offered about the values at the conclusion of execution.

- Register 15 is called the *pc* (program counter).

---

[7]Technically this renders our notation ambiguous. In the Nqthm logic, a decimal number denotes the corresponding natural. Thus, **1101**, for example denotes the natural obtained by taking the one thousand one hundred and first successor of zero. For us to now say that it denotes a list of Booleans is contradictory. But we do not use this informal bit vector notation in formulas. In our English text we make it clear when we have in mind the interpretation of these strings as bit vectors.

7.2.2. The Condition Codes

Recall the four condition code registers of FM9001: carry, overflow, negative, and zero. These can be set and tested by any instruction according to the operation performed by the ALU during that instruction.

The implementation of certain Piton instructions, namely the "test and jump" family and the "arithmetic with carry" family, load and test the condition code registers of FM9001. However, these registers are also treated as temporaries in the machine code.

7.2.3. Boot Code

The first section of the FM9001'memory consists of what is called the *boot code*. The boot code is irrelevant to Piton's implementation; control never enters it and no data from it is read or written. Its purpose was noted at the beginning of Chapter 6. Recall that 'load' takes two arguments in addition to the p-state, *boot-lst* (a list of natural numbers) and *load-addr* (a natural number). The boot code is determined by taking the first *load-addr* elements of *boot-lst*, extending *boot-lst* on the right with 0s if it is not long enough, and converting these *load-addr* natural numbers to bit vectors of length 32 in conventional binary notation. These bit vectors are then loaded into the first *load-addr* memory locations starting at address 0.

7.2.4. User Data Segment

The second partition of the FM9001 memory contains the data segment of the Piton state and is called the *user data segment* of the FM9001 memory. The user data segment is laid out immediately above the boot code. By choosing this arrangement the user of 'load' can specify where the user data segment resides in the FM9001 memory: it starts at address *load-addr*.

The user data segment of the FM9001 machine is isomorphic to the data segment of the Piton machine. Each array in the data segment is allocated a block of words as long as the array. The arrays are laid out successively, each immediately adjacent to its neighbors.

7.2.5. The Program Segment

The third partition, called the *program segment*, holds the binary programs and starts immediately above the user data segment.

The program segment of the FM9001 memory is logically divided into separate areas. Each area corresponds to a Piton program. The contents of such an area is the binary machine code obtained by compiling the corresponding Piton program and then link-assembling it. To compile a Piton program we compile each individual instruction in it, concatenate the results, and sandwich the resulting code between a "prelude" and a "postlude" that implements procedure call and return. We are much more specific later. However, the generated code is easily imagined once we have explained how the abstract stacks and data areas of Piton are concretely implemented.

7.2.6. The System Data Segment

The fourth partition, called the *system data segment*, holds the control and temporary stacks and three words of data used to implement the stack resource instructions and starts immediately above the program segment.

This segment contains five "areas," comparable to the areas (arrays) of the user data segment. In the *control stack area* we represent the Piton control stack, using, in addition, the **cfp** and **csp** registers. In the *temporary stack area* we represent the Piton temporary stack, using, in addition, the **tsp** register. The other three system data areas contain resource bounds used in the implementation of such instructions as **jump-if-temp-stk-empty**.

The amount of space allocated to each stack is one greater than the maximum stack size for that stack, as specified in the Piton state.

Because of the addressing modes in FM9001 it was decided that the stacks should grow "downwards" (i.e., toward absolute address 0). When a push is performed, the stack pointer is decremented and then the operand is deposited into the memory location indicated by the new stack pointer; when a pop is performed, the contents of the indicated memory location is fetched and then the pointer is incremented. When a stack is empty its stack pointer addresses the high word in its area; the contents of this word is never read or written and hence the allocation of the word is actually a waste of one word.[8]

7.2.6.1. The Temporary Stack

Consider the following Piton temporary stack containing four Piton data objects:

$$(obj_1$$
$$obj_2$$
$$obj_3$$
$$obj_4) .$$

obj_1 is the topmost element of the stack. Suppose the maximum temporary stack size is 6.

At the FM9001 level, this temporary stack is represented by a block of 7 words in the system data segment, together with the **tsp** register. Suppose, for concreteness,

---

[8]The implementation could have made the empty stack pointer be an address "just outside" the stack area. However, by allocating the extra word we slightly simplified the correctness proof because we maintained the invariant that the stack pointers were always addresses into the appropriate system data area.

that the temporary stack data area is allocated beginning at absolute address **1000** In Figure 7-2 we show how the above stack is represented. Of course, the contents of the memory locations are not the Piton objects themselves but bit vectors representing them. We deal with this problem later. A push on the stack above would decrement **tsp** and write into address **1001**. A pop would fetch from the current **tsp** and then increment **tsp** to **1003**. If the stack is popped four times it is empty and **tsp** would be **1006**. Because we allocated an extra word to the temporary stack area, the empty stack pointer is still an address into the temporary stack area. Because we will never pop the empty stack, the contents of the extra word is irrelevant.

memory		registers	
address	contents	name	contents
1006	*unused*	**tsp**	1002
1005	obj_4		
1004	obj_3		
1003	obj_2		
1002	obj_1		
1001	*inactive*		
1000	*inactive*		

Figure 7-2: An FM9001 Temporary Stack

7.2.6.2. The Control Stack

The Piton control stack is a stack of frames, each frame containing the bindings for the local variables of the current subroutine and the return program counter. At the FM9001 level, each frame contains the values of the local variables (but not the names) and the return program counter; the machine code references the values by position within the frame. Because the FM9001 frames are just blocks of words laid out consecutively in the control stack area, each frame contains an additional word that points to the "beginning" of the previous frame.

We explain the exact layout of the control stack by example. Below we show a Piton control stack containing three frames.

```
(
  ( ((x . val_x)          ;Frame 1 - current
     (y . val_y)
     (z . val_z))
    ret-pc_1)
  ( ((a . val_a)          ;Frame 2
```

$$(\mathbf{b} \quad . \quad val_b))$$
$$ret\text{-}pc_2)$$
$$(\quad ((\mathbf{u} \quad . \quad val_u) \qquad \text{; Frame 3}$$
$$(\mathbf{v} \quad . \quad val_v))$$
$$ret\text{-}pc_3))$$

Suppose the maximum control stack size is 16. Then the stack above is represented as a block of 17 consecutive words in the system data segment, together with two registers, the **csp** register and the **cfp** register.

For the sake of concreteness, suppose the control stack area begins at address 5000. Then the relevant portion of the FM9001 state is shown in Figure 7-3.

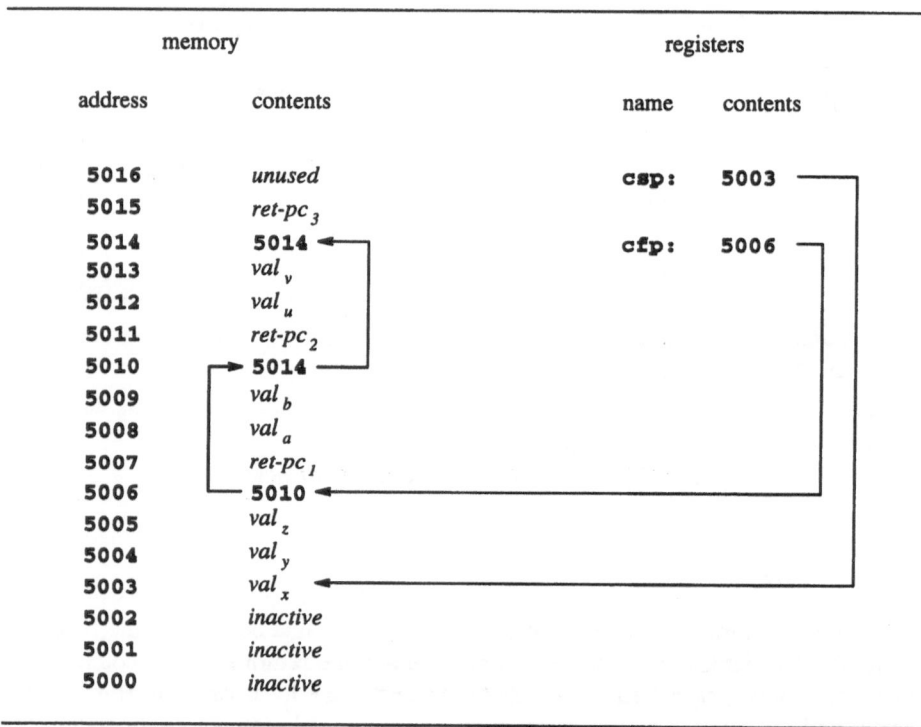

Figure 7-3: An FM9001 Control Stack

Frame 1 of the Piton stack is represented in memory locations **5003-5007** of Figure 7-3. In particular, the return program counter, $ret\text{-}pc_1$, is in "first" word of the top frame (where we enumerate the words in the same order in which they were pushed), location **5007**. The second word of the top frame, location **5006**, contains the address of the frame that was current at the time this frame was built. We call this the *old cfp* word of the frame, for reasons that will become apparent. The bindings are in the last n words of the frame, locations **5003-5005**.

The **csp** register contains **5003**, the "top" of the control stack. The values of the successive local variables are obtained by indexing up from **csp**. The value of the 0th local is at **csp+0**, i.e., **5003**, the value of the 1st local is at **csp+1**, i.e., **5004**, etc.

The **cfp** register contains **5006**. This is the address of the old cfp word in the top frame. That word contains the address of the old cfp word in the previous frame. By starting at **cfp** and threading through the old cfp words one can identify each frame.

Frame 2 of the Piton stack, the one extant when frame 1 was created, is in locations **5008-5011**. Frame 3 is in locations **5012-5015**.

Locations **5000-5002** of the area above are currently unused but will be filled if a new frame is pushed. Location **5016**, the high word in the control stack area, is wasted.

Finally, it should be noted that the deepest frame on the stack, frame 3 in our example, is somewhat pathological. First, its return program counter is irrelevant. This is a peculiarity of Piton, not our implementation. When **ret** is executed in the context of this frame (the top-level entry into the Piton system) the Piton machine halts rather than transfer control out of Piton. So the user of Piton, who starts the machine on some initial state, may select any value for the initial return pc. Second, the old cfp word of that frame is irrelevant for the same reason. In our implementation, the old cfp word of the last frame always points to itself.

7.2.6.3. Stack Resource Limits

In addition to the two stack areas, the system data segment contains three words used to implement the stack resource instructions. These three words each comprise a named area within the system data segment.

The names and contents of the three additional areas are

- *full control stack address*: the address of the lowest word in the control stack area. In our example, it is **5000**. When the control stack pointer is equal to this value, the control stack is full; an additional push would cause the pointer to overflow the control stack area.

- *full temporary stack address*: the address of the lowest word in the temporary stack area. In our example, it is **1000**.

- *empty temporary stack address*: the address of the highest word in the temporary stack area. In our example, it is **1006**. When the temporary stack pointer is equal to this value, the temporary stack is empty.

This completes our sketch of the implementation. In the next four sections we discuss each phase of the implementation in more detail.

7.3. The Intermediate States of Load

Recall that 'load' has four phases and is defined as the composition of four transformations: *r*esource representation (the innermost transformation, 'p⇒r'), compilation into a symbolic form of the FM9001 *i*nstruction set ('r⇒i'), link-assembly into binary *m*achine code ('i⇒m'), and ram tree construction ('m⇒fm9001').[9] In order to compose these transformations it is convenient to define each to return an entire "package" of results to be passed on to the next phase. This is a bit odd since not every transformation changes every component in its package. For example, the resource representation phase, 'p⇒r', translates the stacks of the input p-state into a combination of symbolic registers and linearly allocated arrays, but does not change the program component of the p-state. Nevertheless, the program component of the input p-state is included in the "package" of answers created by 'p⇒r' and that package is passed to the compilation phase. Compilation changes the program component of the package but leaves the stack representations unchanged and passes them down to be linked, etc.

The "packages" are formally represented by new data types in the logic created with the shell principle. The package created by 'p⇒r' is called an "r-state." Similarly, 'r⇒i' creates an "i-state" and 'i⇒m' creates an "m-state." 'M⇒fm9001' creates an FM9001 state. Thus, p-states are transformed into FM9001 states by the composition of 'p⇒r', 'r⇒i', 'i⇒m', and 'm⇒fm9001'. The fact that we call the intermediate results "states" is no accident. To prove the correctness of the downloader we define abstract machines for each of these levels and show that each transformation preserves the semantics of the upper machine on the lower one. For example, in the proof we define the 'r' machine, which operates on r-states. One key step in the proof is to establish a commutative diagram linking the 'p' and 'r' machines via 'p⇒r', roughly that $p⇒r(p(p_0, n))$ is the same as $r(p⇒r(p_0), n)$. We discuss these intermediate abstract machines when we discuss the proof, in Chapter 8. But now that we have explained why we call these intermediate results "states," we will simply regard them as intermediate states in the process of transforming p-states into FM9001 states.

7.4. Resource Representation

Recall that our Piton implementation uses six of the FM9001 registers, to which we have given the names **cfp**, **csp**, **tsp**, **x**, **y**, and **pc**, the four condition code flags, **c**, **v**, **n**, and **z**, and a region of memory in which the system resources are (to be) allocated, called the system data segment. It is the job of 'p⇒r', the first phase of the downloader, to represent Piton's system resources in terms of these FM9001 resources. To do so, 'p⇒r' transforms the input p-state into an "r-state," which is a 15-tuple containing many of the components of the Piton state but in which FM9001 resources are used to represent the stacks and bounds.

---

[9] The italicized letters in the sentence indicate the origins of the letters r, i, and m chosen to name the intermediate phases.

The r-states are a new shell class. The constructor function is 'r-state', which takes 15 arguments. Objects of this class have the following components which we enumerate by the names of the corresponding accessor functions:

- 'r-pc', 'r-cfp', 'r-csp', 'r-tsp', 'r-x', 'r-y': the values of the six registers,

- 'r-c-flg', 'r-v-flg', 'r-n-flg', 'r-z-flg': the values of the four condition code flags,

- 'r-prog-segment': the Piton program segment,

- 'r-usr-data-segment': the Piton data segment,

- 'r-sys-data-segment': the segment devoted to the stacks and maximum stack sizes, and

- 'r-word-size' and 'r-psw': the Piton word size and psw.

The output r-state created by 'p⇒r' is a symbolic description of the use of the FM9001's resources. For example, the value computed for the 'r-pc' "register" is a Piton **pc** object. 'P⇒r' cannot produce absolute addresses for such things as 'r-pc' and 'r-csp' until the link-phase has assigned locations to the various parts of the state.

7.4.1. The System Data Segment and System Addresses

The 'r-sys-data-segment' of an r-state contains the symbolic description of the system data segment. That description is a list structure similar to Piton's data segment. In particular, the system data segment is described by a list of length five. Each element describes a *system data area* by listing the *name* of the area and the *array* of objects associated with it. We give the following literal atom names to the areas in the system data segment: **cstk**, **tstk**, **full-ctrl-stk-addr**, **full-temp-stk-addr**, and **empty-temp-stk-addr**.

Intuitively, the array associated with each area is just a list of tagged Piton objects. This is not completely accurate because it is necessary for the control stack to contain addresses into itself, namely the old cfp words, and no Piton data object addresses the system data segment. We therefore introduce an eighth type of object, called a *system data address*, which is a pair of the form (*name* . *n*), where *name* is one of the area names above and *n* is a natural number less than the length of the associated area.[10] Thus, (**full-ctrl-stk-addr** . 0) is the symbolic address of the full control stack address word in the system data segment. Recall that that word contains the address of the lowest word in the control stack area. In fact we can now write that address down symbolically too. It is (**cstk** . 0).

System addresses will be tagged with the literal atom **sys-addr**, exactly

---

[10]The implied motivation for system addresses was to represent the old cfp words in the control stack. We have implemented a general purpose system addressing notation because system addresses are used throughout the implementation. For example, they may be found in the compiler output.

analogous to the way data addresses are tagged with **addr**. Thus, if we write
(sys-addr (cstk . 25)) we are referring to the 25th word in the control
stack area in the system data segment. If we write **(addr (cstk . 25))** we are
referring to the 25th word in a global data area named **cstk** (supposing one exists)
in the Piton data segment. We usually omit the tag in informal text and simply say
"the system address **(cstk . 25)**" or "the data address **(cstk . 25)**."

The arrays associated with each of the five area names are simply lists of either
Piton objects or system addresses. The previous section sketching the layout of the
stacks should make it clear how we load these arrays.

We conclude with a simple example. Suppose the maximum control stack size is
11 and the maximum temporary stack size is 8. Suppose the control stack of the
Piton state is

```
((((x . (nat 0))           ; Frame 1 - current
   (y . (nat 1))
   (z . (nat 2)))
  (pc (main . 4)))
 (((a . (int 1))           ; Frame 2
   (b . (int 2)))
  (pc (main . 0))))).
```

Suppose the temporary stack is

```
((addr (a . 3))            ; topmost element
 (bool t)
 (nat 27)).
```

The symbolic description of the corresponding system data segment is shown in
Figure 7-4.

7.4.2. The Registers

The resource representation phase also specifies the symbolic value of each of the
registers. This is necessary since the registers participate in the representation of the
stacks. The **pc** register is exactly the same as in the Piton state. The **cfp**, **csp** and
tsp registers are set to the appropriate system addresses. For the system data
segment in Figure 7-4 the register values are

```
cfp:       (sys-addr (cstk . 4))
csp:       (sys-addr (cstk . 1))
tsp:       (sys-addr (tstk . 5))
```

The initial values of the **x** and **y** registers and of the condition code flags are
irrelevant.

The system addresses, along with all the other data objects in the resource
representation description, will be turned into bit vectors by the link phase, when we
know where, in absolute terms, the system data segment will be located.

	; offset	comment
`((cstk (nat 0)`	; 0	inactive
`     (nat 0)`	; 1	value of **x** - Frame 1
`     (nat 1)`	; 2	value of **y**
`     (nat 2)`	; 3	value of **z**
`     (sys-addr (cstk . 8));`	; 4	old cfp
`     (pc (main . 4))`	; 5	return pc
`     (int 1)`	; 6	value of **a** - Frame 2
`     (int 2)`	; 7	value of **b**
`     (sys-addr (cstk . 8));`	; 8	old cfp
`     (pc (main . 0))`	; 9	return pc
`     (nat 0))`	;10	unused
`  (tstk (nat 0)`	; 0	inactive
`     (nat 0)`	; 1	inactive
`     (nat 0)`	; 2	inactive
`     (nat 0)`	; 3	inactive
`     (nat 0)`	; 4	inactive
`     (addr (a . 3))`	; 5	topmost element
`     (bool t)`	; 6	
`     (nat 27)`	; 7	btm-most element
`     (nat 0))`	; 8	unused
`(full-ctrl-stk-addr (sys-addr (cstk . 0)))`		
`(full-temp-stk-addr (sys-addr (tstk . 0)))`		
`(empty-temp-stk-addr (sys-addr (tstk . 8))))`		

Figure 7-4: A System Data Segment

7.5. Compiling

The next phase of the 'load' process, 'r⇒i', is compilation. The input is an r-state and the output is an "i-state."

The i-states are a new shell class, constructed by the function 'i-state' of 15 arguments. The components of an i-state correspond closely with those of r-states and have analogous names:

- 'i-pc', 'i-cfp', 'i-csp', 'i-tsp', 'i-x', 'i-y',

- 'i-c-flg', 'i-v-flg', 'i-n-flg', 'i-z-flg',

- 'i-prog-segment', 'i-usr-data-segment', 'i-sys-data-segment',

- 'i-word-size', and 'i-psw'.

When an r-state is transformed into an i-state, all the components except the 'r-pc' and 'r-prog-segment' are left unchanged (i.e., they are moved into the corresponding components of the output i-state). The 'i-prog-segment' of the i-state is obtained by compiling the programs in the r-state. The 'i-pc' is correspondingly mapped into a program counter into the new program segment.

The compiler scans the program segment of the input r-state (which is just the program segment of the original Piton state) and pairs each program name with the assembly code for that program. The "assembly code" generated by the 'r⇒i' compiler is what we "i-code." I-code is a special-purpose, symbolic, annotated FM9001 machine code. I-code is special-purpose because it only supports the resources used by Piton (e.g., it only has names for six registers) and its instructions and annotations support Piton's data types. I-code instructions translate 1:1 into FM9001 machine code. We discuss i-code at much greater length later. The compilation of each Piton program into i-code is independent of the other programs and of the other components of the Piton state. It is the link phase that worries about references between programs and data.

The assembly code for a Piton program is logically divided into three parts. The first part, called the *prelude*, is executed as part of subroutine **call**. The prelude builds the new control stack frame for the invocation, removes the actuals from the temporary stack and stores them in the newly built control stack frame. The second part, called the *body*, is the translation of the successive instructions in the body of the Piton program. The third part, called the *postlude*, is executed as part of the **ret** instruction and pops the top frame off the control stack, returning control to the indicated pc.

We exhibit a simple compilation below. Consider the following silly Piton program:

```
(demo (x y z)                    ; Formals x, y, and z
      ((a (int -1))              ; Temporaries a and i
       (i (nat 2)))
      (push-local y)             ; pc 0
      (push-constant (nat 4))    ; pc 1
      (add-nat)                  ; pc 2
      (ret))                     ; pc 3
```

The program, named **demo**, has three formals and two locals. It adds 4 to the value of the second formal and leaves the result on the stack.

The output of the compiler (actually of the function 'icompile-program') on this program is shown in Figure 7-5. The output a list constant and is printed according to our usual rules for displaying list constants but we have formatted it by adding "whitespace" so that it appears in a three column format to make its structure clearer. The first column consists entirely of labels and i-code instructions.[11] Note that our symbolic code uses the same **dl** construct used by Piton to attach a label and a comment to an instruction. The second column contains the Piton source code used as input to the compiler. The compiler uses the comment field of the **dl** construct to annotate each block of i-code with the high level instruction that generated it. The third column consists of manually inserted comments that enumerate the successive values of the i-code program counter.

---

[11]FM9001 machine code instructions can take "immediate" data from the next word in the instruction stream using the post-increment addressing mode with the program counter register. In our i-code, such double-word instructions have names ending in a "*."

i-code labels and instructions	Piton instruction	pc
(demo		
(dl (demo prelude)	(prelude)	
(cpush_cfp))		; 0
(move_cfp_csp)		; 1
(cpush_*)		; 2
(nat 2)		; 3
(cpush_*)		; 4
(int -1)		; 5
(cpush_<tsp>+)		; 6
(cpush_<tsp>+)		; 7
(cpush_<tsp>+)		; 8
(dl (demo . 0)	(push-local y)	
(move_x_*))		; 9
(nat 1)		;10
(add_x{n}_csp)		;11
(tpush_<x{s}>)		;12
(dl (demo . 1)	(push-constant (nat 4))	
(tpush_*))		;13
(nat 4)		;14
(dl (demo . 2)	(add-nat)	
(tpop_x))		;15
(add_<tsp>{n}_x{n})		;16
(dl (demo . 3)	(ret)	
(jump_*))		;17
(pc (demo . 4))		;18
(dl (demo . 4)	(postlude)	
(move_csp_cfp))		;19
(cpop_cfp)		;20
(cpop_pc))		;21

Figure 7-5: Compiler Output for **demo**

Observe that the code for **demo** in Figure 7-5 can be broken down into blocks by the **dl** labels. Each such block of i-code is called a *basic block*. For example, the basic block of the prelude extends from **pc 0** through **pc 8**; the basic block of (**push-local y**) extends from **9-12**. Each basic block starts with the definition of a label. For example, the prelude is labeled by (**demo prelude**), a list object in the logic and guaranteed to be unique among the labels used by the assembler. All labels, of course, are eventually removed by the linker and merely serve as unique entries in the link table.

Consider the prelude. It builds a new frame on the control stack. The construction of the frame begins before the prelude is entered, when the **call** instruction is executed. **Call** pushes the return program counter onto the control stack and jumps to the prelude of the called subroutine. Recall that the first word in the new frame is the return program counter (which has just been pushed). The prelude builds the rest of the frame by pushing more words onto the control stack.

Below we display the basic block of the prelude for **demo**.

```
(cpush_cfp)                                          ; 0
(move_cfp_csp)                                       ; 1
(cpush_*)                                            ; 2
(nat 2)                                              ; 3
(cpush_*)                                            ; 4
(int -1)                                             ; 5
(cpush_<tsp>+)                                       ; 6
(cpush_<tsp>+)                                       ; 7
(cpush_<tsp>+)                                       ; 8
```

At instruction **0** the current frame pointer register, **cfp**, is pushed onto the control stack. That sets the old cfp word of the frame. At instruction **1** the current **csp**, which now points to the old cfp word just pushed, is moved into the **cfp** register. Thus, the **cfp** register points to the old cfp word of the frame under construction. The instruction at **2** is a double-word instruction that pushes the natural number **2** onto the control stack. This is the initial value of the temporary variable **i**. At instruction **4** the initial value of the temporary variable **a** is pushed. The last three instructions of the prelude, **6-8**, each pop one thing off the temporary stack and push the result onto the control stack. These instructions move the actuals into the new frame. This completes the construction of the frame. Note that the **csp** register points to the top of the stack and the **cfp** register points to the old cfp word as required.

The first executable Piton instruction in the program, **(push-local y)**, is compiled as instructions **9-12**. That basic block is labeled in the assembly language by **(demo . 0)** which happens to be the Piton program address of the Piton instruction for which the code was generated. *Each program address at the Piton level is defined as a label in the i-code.* It is via this identification of Piton program address objects with i-code labels that the linker is able to replace each **pc** object by its absolute address. However, there may be objects tagged **pc** in our i-code that are not legal program address objects in Piton, e.g., **(demo prelude)**. In general, the tag **pc** in i-code means "i-code label," not Piton program counter.

The instructions for **(push-local y)** are

```
(move_x_*)                                           ; 9
(nat 1)                                              ;10
(add_x{n}_csp)                                       ;11
(tpush_<x{s}>)                                       ;12
```

The basic idea is to fetch a certain element from the top frame and push it onto the temporary stack. The element is the one in the slot for the local variable **y**, which is in position **1** of the locals. Recall that the locals are numbered from **0** and **x** is thus the 0th local of the program; the value of **x** is thus at the address indicated by **csp**. The value of **y** is at the address one greater than **csp**. The code above may be explained as follows: **(9)** Move the index, **1**, of the local variable into the **x** register. **(11)** Add the contents of the **csp** register to **x** and store the result in **x**; this is the address of the appropriate slot in the control stack. **(12)** Fetch indirect through the **x** register and push the result onto the temporary stack.

In our symbolic code, we use angle brackets, e.g., **<x>**, around a register to indicate register-indirect addressing mode. We use set braces, e.g., **{s}**, to indicate the type of object in the register. Thus, the instruction **add_x{n}_csp** means "add the contents of **csp** to *the natural number in* **x**" and **tpush_<x{s}>** means "indirect through *the system data address in* **x**." We discuss the data type annotations later.

The next instruction in the Piton program is **(push-constant (nat 4))** and the code generated is

```
(tpush_*)                                        ;13
(nat 4)                                          ;14
```

This code pushes the natural number 4 onto the temporary stack.

The next Piton instruction is **(add-nat)**, which is supposed to pop two naturals off the temporary stack, add them together, and push the result. The generated code is

```
(tpop_x)                                         ;15
(add_<tsp>{n}_x{n})                              ;16
```

The code pops one thing off the temporary stack into **x**. Then it adds the natural in **x** to the natural fetched indirect through **tsp** (the top item on the stack), and deposits the result indirect through **tsp** (back onto the top of the stack).

The last executable instruction in the Piton code is the return instruction. It compiles into

```
(jump_*)                                         ;17
(pc (demo . 4))                                  ;18
```

This is just an unconditional jump to the label **(demo . 4)**, which is where the postlude is located.[12]

The postlude is

```
(move_csp_cfp)                                    ;19
(cpop_cfp)                                        ;20
(cpop_pc)                                         ;21
```

The postlude must remove the top frame from the stack, restore the **cfp** register to the value it had at the time of the **call**, and restore the program counter to the return pc. The first instruction moves the contents of the **cfp** register into **csp**. This effectively pops all the bindings of this frame and makes the top of the stack be the old cfp word. The next instruction pops the control stack into the **cfp** register,

---

[12]Our compiler has much room for improvement. The instruction at **17** jumps to the next executable instruction, and so could be eliminated. We do not do any such optimizations.

restoring **cfp** and exposing the return pc at the top of the stack. The last instruction pops the control stack into the **pc**, completing the return and removing the last vestige of the now popped frame.

Now consider the following segment of a Piton program **main** that calls **demo** on the address of the 25th element of the array **delta1**, the natural number 17, and the Boolean value **t**. Suppose the **call** instruction is located at program address **(main . 3)**.

```
(push-constant (addr (delta1 . 25)))
(push-constant (nat 17))
(push-constant (bool t))
(call demo)
```

The i-code generated for this segment is shown below. We have stripped out the label definitions.

```
(tpush_*)                    ;Push first actual on temp
(addr (delta1 . 25))
(tpush_*)                    ;Push second actual on temp
(nat 17)
(tpush_*)                    ;Push third actual on temp
(bool t)
(cpush_*)                    ;Push return pc on ctrl
(pc (main . 4))
(jump_*)                     ;Jump to (demo prelude)
(pc (demo prelude))
```

Observe that all addresses, both program and data, are represented in the i-code by *extended data objects*—i.e., either Piton data objects, system data address objects, or i-code labels tagged **pc**. This is true regardless of how the address originated. For example, **(delta1 . 25)** was originally a data object in the source program. The reference to **demo** occurred in the **call** instruction and has been transformed into the i-code data object **(demo prelude)** of type **pc**. The return pc **(main . 4)** above was only implicit in the source program.

The next example compilation illustrates our handling of labels. Consider the following program, **ptz** ("pop till zero"), which pops the temporary stack until it pops a **0**.

```
(ptz nil
     nil
  (dl loop ()
      (test-nat-and-jump zero end))
      (jump loop)
  (dl end ()
      (ret)))
```

In Figure 7-6 we show the compiler output for **ptz**.

Let us look carefully at the code generated for the **test-nat-and-jump** instruction. It is supposed to pop the stack and jump to the label **end** if the result is the natural number 0. The i-code is in locations **2-5**. The first instruction pops the

i-code labels and instructions	Piton instruction	pc
`(ptz`		
`(dl (ptz prelude)`	`(prelude)`	
`(cpush_cfp))`		`; 0`
`(move_cfp_csp)`		`; 1`
`(dl (ptz . 0)`	`(dl loop nil`	
	`(test-nat-and-jump zero`	
	`end))`	
`(tpop{n}_<z>_y))`		`; 2`
`(move_x_*)`		`; 3`
`(pc (ptz . 2))`		`; 4`
`(jump-z_x)`		`; 5`
`(dl (ptz . 1)`	`(jump loop)`	
`(jump_*))`		`; 6`
`(pc (ptz . 0))`		`; 7`
`(dl (ptz . 2)`	`(dl end nil (ret))`	
`(jump_*))`		`; 8`
`(pc (ptz . 3))`		`; 9`
`(dl (ptz . 3)`	`(postlude)`	
`(move_csp_cfp))`		`;10`
`(cpop_cfp)`		`;11`
`(cpop_pc))`		`;12`

Figure 7-6: Compiler Output for **ptz**

stack into the **y** register and sets the **z** condition code (according to whether the result is 0). Next, we move into the **x** register the **pc** data object (**ptz . 2**). Then we jump to the contents of **x** if the **z** condition code is set. Inspection will show that (**ptz . 2**) is the i-code label marking the point labeled **end** in the Piton source code.

Similarly, observe the compilation of the (**jump loop**) instruction. The i-code is at locations **6-7** above. It reads: Jump to i-code label (**ptz . 0**).

In general, references to labels in Piton are compiled into references to **pc** type data objects.

As noted above, all program and data addresses, whether implicit or explicit are explicitly mentioned extended data objects (of type **pc**, **addr**, **sys-addr**, or **subr**) in the i-code produced by the Piton compiler. It is the job of the linker, discussed next, to replace these objects by the corresponding absolute addresses. Until this is done, the code and the data segment are relocatable.

Recall that we have introduced a new type of object, the system address, which exists in our implementation of Piton but is not one of the types in Piton. When the compiler must make a reference to a word in the system data segment it uses these symbolic addresses so that the code is relocatable with respect to where the system data segment is laid out.

To illustrate the use of these internal addresses, consider the compilation of

```
(jump-if-temp-stk-full error)
```

and suppose that the label **error** is defined at **pc (main . 152)**. The i-code
generated for this instruction is

```
(move_x_tsp)                              ;0
(move_y_*)                                ;1
(sys-addr (full-temp-stk-addr . 0))      ;2
(move_y_<y{s}>)                          ;3
(sub_<z>_x{s}_y{s})                      ;4
(move_x_*)                                ;5
(pc (main . 152))                         ;6
(jump-z_x)                                ;7
```

The code first puts the temporary stack pointer into the **x** register so we can do
some arithmetic on it. Then, in lines **1-3** above, the code fetches into the **y** register
the address of the first word in the temporary stack area. This is done in two
instructions. First (in lines **1-2**), we load into **y** the system address
(full-temp-stk-addr . 0). In our implementation, the contents of this address is the address of the first word in the temporary stack area. On line **3** we fetch
indirect through the system address in **y** and put the result in **y**. This loads **y** with
the address of the first location on the temporary stack area.[13]

On line **4** we subtract the system address in **y** from the system address in **x** and set
the **z** condition code register. Thus, the **z** register is **t** if and only if **x** and **y** were
equal. On lines **5-6** we move into **x** the address to which we wish to jump. On line
7 we jump if **z** is true.

The examples shown in this section are fairly representative of the code generated
by the Piton compiler. A complete listing of the compiler is given in Appendix IV.

7.6. The Link-Assembler

Traditionally, compilers produce relocatable assembly code, which is then turned
into relocatable machine code by an ''assembler'', and then into absolute machine
code by a ''linker.'' In addition, assemblers and linkers must traditionally consider
the user's data declarations and initialization too. We do not follow this paradigm
rigidly but the basic concepts are still present in our ''link-assembler.''

By the time the link-assembler is invoked the first two phases of 'load' have been
carried out: the resource representation phase has produced an r-state containing
symbolic descriptions of the registers and the system data segment; the compiler has

---

[13]In fact, the code for **jump-if-temp-stk-full** could be shortened by eliminating the indirection
through **full-temp-stk-addr** and simply loading **(sys-addr (tstk . 0))** into the **y** register.
We introduced **full-temp-stk-addr** primarily to force our proof to deal with such common
implementation invariants as ''the contents of this fixed address is the address of this more fluid
boundary.''

produced from that an i-state containing the i-code version of the program segment and pc. The user data segment of the i-state is just that of the original p-state. The job of our "link-assembler," 'i⇒m', is to replace the symbolic instructions and data objects in the i-state by concrete bit vectors. The link-assembler produces an "m-state" as its output.

The m-states are a new shell class constructed by 'm-state'. M-states are objects containing 6 components. They are

- 'm-regs': a list of 16 registers, each of which is a bit vector of length **32**. Ten of the 16 are set to the **0**-vector. The other six contain the linked values of the six register components of the input i-state.

- 'm-c-flg', 'm-v-flg', 'm-n-flg', 'm-z-flg': Boolean valued flags corresponding to the four flags of the i-state.

- 'm-mem': a list of 32-bit long bit vectors containing the linked values of the boot code, the user data segment, the program data segment, and the system data segment.

To construct these lists of bit vectors, the link-assembler first builds a collection of "link tables" indicating where each program, label, system data area and user data area are located in absolute terms. This is done by the function 'i-link-tables' which is defined on page 272. Note that by the time we build the link tables, the user specified boot code and load address are available and the three Piton-specific segments of the FM9001 memory are symbolically described by the Piton data segment, the i-code program segment, and the system data segment. All three of these symbolic descriptions have the same form: each is a list of pairs consisting of the name of the program or area and an array listing the contents of the area. Each element of the array is either an (optionally labeled) i-code instruction or an extended data object. Each element can be mapped to a single word in FM9001. Thus, the number of words to be allocated to each area is just the length of the associated array. The absolute location of each name can be determined by summing the lengths of the areas preceding the definition of the name. The absolute location of each label can be similarly determined by counting the number of items in each i-code program preceding the label definition.

Once the link tables have been created, the link-assembler scans each of the three Piton memory segments in turn. Each i-code instruction is replaced by the corresponding FM9001 machine code instruction, by a function which can be thought of as the basic component of an assembler. Each data object is replaced by the corresponding bit vector, using the link tables as appropriate.

We present the link-assembler in three subsections. First, we describe how individual i-code instructions are assembled into FM9001 machine code instructions. Then we discuss how we generate the link tables. Finally, we describe how we transform each of the data objects.

7.6.1. The Instruction Assembler

The instruction assembler converts a single i-code instruction into an FM9001 machine code instruction, i.e., a 32-bit wide bit vector. The conversion is done in two steps and explicitly involves an assembly language for FM9001. In essence, each i-code instruction is taken as a "pseudo-instruction" that is mapped first into an assembly instruction and then into a bit vector. The formalization of this concept is embodied in the function 'link-instr-word' which is defined on page 289.

7.6.1.1. Expanding I-code into Assembly Code

Each i-code instruction is of the form (**opcode**), where **opcode** is a literal atom in the logic and completely describes the instruction. Our i-code instructions do not have operands, even though their names e.g., **add_<tsp>{n}_x{n}**, suggest more structure. We explain the design of i-code when we discuss the proof of the correctness of our implementation.

An *association list*, or The following table, which is an "association list," is the map from i-code opcodes to assembly instructions. *alist*, is a list of pairs representing a table; each pair is an entry in the table, showing a key (in the 'car') and its associated value (in the 'cdr' or, as below, 'cadr'). In the table shown below, keys include all of the i-code opcodes and the values are the associated FM9001 assembly code instructions.

Map from I-code to Assembly Code

i-code	assembly code
((add_<c>_x_x{n}	(add (c) x x))
(add_<tsp>_<tsp>{v}	(add () (tsp) (tsp)))
(add_<tsp>_<tsp>{n}	(add () (tsp) (tsp)))
(add_<tsp>{a}_x{n}	(add () (tsp) x))
(add_tsp_*{n}	(add () tsp (pc +1)))
(add_tsp_x{n}	(add () tsp x))
(add_<tsp>{i}_x{i}	(add () (tsp) x))
(add_<tsp>{n}_x{n}	(add () (tsp) x))
(add_pc_x{n}	(add () pc x))
(add_x_x{n}	(add () x x))
(add_x{n}_csp	(add () x csp))
(addc_<c>_x{n}_y{n}	(addc (c) x y))
(addc_<v>_x{i}_y{i}	(addc (v) x y))
(and_<tsp>{v}_x{v}	(and () (tsp) x))
(and_<tsp>{b}_x{b}	(and () (tsp) x))
(asr_<c>_<tsp>_<tsp>{b}	(asr (c) (tsp) (tsp)))
(cpop_cfp	(move () cfp (csp +1)))
(cpop_pc	(move () pc (csp +1)))
(cpush_*	(move () (-1 csp) (pc +1)))
(cpush_<tsp>+	(move () (-1 csp) (tsp +1)))
(cpush_cfp	(move () (-1 csp) cfp))
(decr_<tsp>_<tsp>{i}	(decr () (tsp) (tsp)))
(decr_<tsp>_<tsp>{n}	(decr () (tsp) (tsp)))
(incr_<tsp>_<tsp>{i}	(incr () (tsp) (tsp)))
(incr_<tsp>_<tsp>{n}	(incr () (tsp) (tsp)))
(incr_y_y{n}	(incr () y y))
(int-to-nat	(move () x x))
(jump-n_x	(move-n () pc x))
(jump-nn_x	(move-nn () pc x))
(jump-nz_x	(move-nz () pc x))
(jump-z_x	(move-z () pc x))
(jump_*	(move () pc (pc)))
(jump_x{subr}	(move () pc x))
(lsr_<c>_x_x{n}	(lsr (c) x x))
(lsr_<tsp>_<tsp>{v}	(lsr () (tsp) (tsp)))
(move-c_<tsp>_*	(move-c () (tsp) (pc +1)))
(move-v_<tsp>_*	(move-v () (tsp) (pc +1)))
(move-z_<tsp>_*	(move-z () (tsp) (pc +1)))
(move-n_x_*	(move-n () x (pc +1)))
(move_<tsp>_*	(move () (tsp) (pc +1)))
(move_<x{a}>_<tsp>	(move () (x) (tsp)))
(move_<x{s}>_<tsp>	(move () (x) (tsp)))
(move_cfp_csp	(move () cfp csp))
(move_csp_cfp	(move () csp cfp))
(move_x_*	(move () x (pc +1)))

```
(move_x_<x{s}>              (move () x (x)))
(move_x_tsp                 (move () x tsp))
(move_x_x                   (move () x x))
(move_y_*                   (move () y (pc +1)))
(move_y_<y{s}>              (move () y (y)))
(move_y_tsp                 (move () y tsp))
(neg_<tsp>_<tsp>{i}         (neg () (tsp) (tsp)))
(not_<tsp>_<tsp>{v}         (not () (tsp) (tsp)))
(or_<tsp>{v}_x{v}           (or () (tsp) x))
(or_<tsp>{b}_x{b}           (or () (tsp) x))
(sub_<c>_<tsp>{a}_x{a}      (sub (c) (tsp) x))
(sub_<c>_<tsp>{n}_x{n}      (sub (c) (tsp) x))
(sub_<nv>_<tsp>{i}_x{i}     (sub (n v) (tsp) x))
(sub_<tsp>{a}_x{n}          (sub () (tsp) x))
(sub_x{s}_y{n}              (sub () x y))
(sub_<tsp>{i}_x{i}          (sub () (tsp) x))
(sub_<tsp>{n}_x{n}          (sub () (tsp) x))
(sub_<tsp>{s}_x{s}          (sub () (tsp) x))
(sub_<z>_x{s}_y{s}          (sub (z) x y))
(subb_<c>_x{n}_y{n}         (subb (c) x y))
(subb_<v>_x{i}_y{i}         (subb (v) x y))
(tpop_<c>_x                 (move (c) x (tsp +1)))
(tpop_<x{a}>                (move () (x) (tsp +1)))
(tpop_<x{s}>                (move () (x) (tsp +1)))
(tpop_pc                    (move () pc (tsp +1)))
(tpop_x                     (move () x (tsp +1)))
(tpop_y                     (move () y (tsp +1)))
(tpop{v}_<z>_y              (move (z) y (tsp +1)))
(tpop{b}_<z>_y              (move (z) y (tsp +1)))
(tpop{i}_<zn>_y             (move (z n) y (tsp +1)))
(tpop{n}_<z>_y              (move (z) y (tsp +1)))
(tpush_*                    (move () (-1 tsp) (pc +1)))
(tpush_<x{a}>               (move () (-1 tsp) (x)))
(tpush_<x{s}>               (move () (-1 tsp) (x)))
(tpush_csp                  (move () (-1 tsp) csp))
(tpush_tsp                  (move () (-1 tsp) tsp))
(tpush_x                    (move () (-1 tsp) x))
(xor_<tsp>_<tsp>            (xor () (tsp) (tsp)))
(xor_<tsp>{v}_x{v}          (xor () (tsp) x))
(xor_<tsp>{b}_*{b}          (xor () (tsp) (pc +1)))
(xor_<tsp>{b}_x{b}          (xor () (tsp) x))
(xor_<z>_<tsp>_x            (xor (z) (tsp) x)))
```

There are a total of 87 i-code opcodes. But some distinct i-code opcodes map to the same assembly language instruction. For example, both xor_<tsp>{v}_x{v} and xor_<tsp>{b}_x{b} map to (xor () (tsp) x). As is suggested by the annotations "{v}" and "{b}" in the opcode names, the first instruction deals with bit vectors and the second deals with Booleans. As manifested by the fact that they both map to a single instruction, no such distinction exists at the concrete level of FM9001. Why then do we have two different i-code instructions? The answer, for the moment, is that the type annotations serve the useful mnemonic role of helping us keep straight the types of objects we are manipulating. However, as we explain when we discuss the correctness proof, the annotations play a much deeper role: they let us factor the proof into a compiler proof and a link-assembler proof.

7.6.1.2. The Assembly Language

There are only 69 distinct assembly language instructions in the i-code table above. These could have been converted into the corresponding bit vectors by hand and listed in the table. However, it was less error-prone to implement a general purpose assembler for FM9001.

The structure of the assembly language should be pretty obvious from the examples above. The form of an assembly instruction is

$(op \ cc \ b \ a)$,

where

- *op* determines the operation code of the FM9001 instruction and is any of following literal atoms **incr, addc, add, neg, decr, subb, sub, ror, asr, lsr, xor, or, and, not, move, move-nc, move-c, move-nv, move-v, move-nz, move-z, move-nn** or **move-n**;

- *cc* determines which of the condition code registers are set by the instruction and is a list containing any subset of {**c, v, n, z**}; and

- *b* and *a* determine the two operands of the instruction and are each of one of the following forms: *reg*, (*reg*), (**-1** *reg*) or (*reg* **+1**) where *reg* is one of the following literal atoms: **pc, cfp, csp, tsp, x** or **y**.

The four different forms of operands describe both a register and an address mode. The address modes described are, respectively, register direct, register indirect, register indirect with pre-decrement, and register indirect with post-increment.

To assemble an FM9001 instruction, e.g., 32-bit wide bit vector, from an assembly instruction, the instruction assembler uses various tables to map the literal atoms above to numbers. For example, the **lsr** operation code is mapped to the decimal number 10, which in binary is **1010**, the FM9001 opcode for the "logical shift right, top bit zero" instruction. The condition codes and register names are similarly mapped to particular bit positions and register numbers. The FM9001 instruction is then assembled by arithmetic and finally converted to a bit vector. The details are given in the definition of **mci**, page 293.

7.6.1.3. An Example

Consider the i-code instruction (**tpush_x**), which pushes the contents of the **x** register onto the temporary stack in **tsp**. This instruction is mapped by the i-code table into the assembly instruction (**move () (-1 tsp) x**). This in turn is assembled into the natural number **266374148**, which when converted to a bit vector in standard 32-bit binary notation is

00001111111000001000110000000100.

Decoding this vector as an FM9001 instruction yields

```
0000  1111  1110  0000  10  0011  0  000  00  0100
  |     |     |     |    |    |    |         |  | reg A: x
  |     |     |     |    |    |    |         | mode A: direct
  |     |     |     |    |    |    |     not immediate
  |     |     |     |    |    | reg B: tsp
  |     |     |     |    | mode B: pre-decrement
  |     |     |     | set no condition codes
  |     |   store code: T
 MOVE
```

That is, the instruction is an unconditional (store code **1110**) move (*op*: **1111**). Operand A, the source of the move, is register 4 (**0100**), otherwise known as the **x** register, in register-direct mode (**00**). Operand B, the destination of the move, is register 3 (**0011**), the **tsp** register, in register-indirect with pre-decrement mode (**10**). The effect of the instruction is to (a) increment the program counter, register 15, by 1; (b) fetch the contents, *x*, of register 4; (c) decrement the contents of register 3 by 1 and store the result in 3; (d) deposit *x* at the address contained in register 3. Observe that if register 3 contains a stack pointer and one uses the conventions that the stack pointer points to the topmost element of the stack and stacks grow downward, then this instruction pushes *x* onto the stack pointed to by register 3. That is, it pushes the contents of the **x** register onto the stack indicated by the **tsp** register.

7.6.1.4. Use of the Addressing Modes

As the i-code table shows, the assembler makes extensive use of all four addressing modes. The effects achieved with the various combinations of modes is sometimes subtle. We explain our use of the modes here.

Pushes onto the stacks are achieved by using pre-decrement addressing mode on the stack pointer. Thus, as we saw above, (**tpush_x**) — which pushes the **x** register onto the temporary stack—is implemented by (**move () (-1 tsp) x**).

Pops are achieved by using post-increment addressing mode. Thus, (**tpop_x**) — which pops the top of the temporary stack into the **x** register—is implemented with (**move () x (tsp +1)**). The effect of this instruction is to

(a) increment the program counter by 1; (b) fetch operand A indirect through the address in the **tsp** register (obtaining the topmost element of the stack) ; (c) increment the **tsp** register by 1 (popping the stack); and (d) deposit operand A into the **x** register.

When we move the actuals from the temporary stack to the control stack we can move each with one instruction: **(cpush_<tsp>+)** which maps to **(move ()** **(-1 csp) (tsp +1))**. This instruction fetches the topmost element of **tsp**, pops **tsp**, and pushes the operand fetched onto **csp**.

Perhaps the most confusing addressing mode combination is when we use post-increment mode on the program counter.   Consider the i-code instruction **(tpush_*)**. This instruction is supposed to push onto the temporary stack the next word in the instruction stream. Thus, the i-code sequence

```
(tpush_*)
(nat 27)
(add_tsp_x{n})
```

will push 27 onto the temporary stack and then execute the **(add_tsp_x{n})** instruction. **(Tpush_*)** is mapped into **(move () (-1 tsp) (pc +1))**. The effect of this instruction is to (a) increment the program counter by 1, as usual, so that the **pc** now points to the **(NAT 27)**; (b) fetch operand A indirect through the **pc** register, which causes operand A to be **27**; (c) (post-)increment the **pc** register by 1, which causes the **pc** to point to the **add_tsp_x{n}** instruction; and (d) push operand A onto the temporary stack by (pre-)decrementing **tsp** by 1 and storing operand A into the resulting address.

This completes our discussion of how i-code instructions are assembled into FM9001 machine code instructions. The details can be gleaned by reading the definition of 'link-instr-word', defined on page 289.

7.6.2. The Link Tables

We use four link tables. The first, called the *program link table*, maps each program name to the absolute location of the beginning of the program. The second table, called the *label tables*, maps each each program name to its "label table." The *label table* for a program name maps the i-code labels to the absolute position of the definition of the label in the i-code program. We illustrate this in a moment. The third table, called the *user data link table*, maps each global data area name to the absolute location of the beginning of the associated array. The fourth table, called the *system data link table*, maps each of the five system data area names to the absolute location of the beginning of the associated area.

Here is a simple example. Consider the Piton data segment

```
((a        (nat  0)
           (nat  1)
           (nat  2)
           (nat  3)
           (nat  4))
  (b       (int  -2)
           (int  -1)
           (int  0))
  (a-len (nat  5))
  (b-len (nat  3)))
```

Observe that this data segment requires a total of 10 words: 5 are in the **a** area, 3 in the **b**, and 1 each in the remaining two. If we were to download (a Piton state containing) this data segment into *load-addr* **8000**, then the first word of the **a** area would go into FM9001 memory location **8000**. The user data link table would thus be

```
((a        .  8000)
  (b        .  8005)
  (a-len .  8008)
  (b-len .  8009)).
```

Thus, the 0th word of the array **b** is at absolute location **8005**.

Suppose the Piton state in question contained the following system of programs.

```
((main nil nil
          (call ptz)
          (ret))
  (ptz   nil nil
    (dl   loop ()
          (test-nat-and-jump zero end))
          (jump loop)
    (dl   end   ()
          (ret)))).
```

The i-code produced by compiling the above system is shown in Figure 7-7. Observe that **main** contains 11 FM9001 instructions and **ptz** contains 13.

The program link table produced from the i-code is

```
((main  .  8010)
  (ptz    .  8021)).
```

That is, the **main** program is loaded starting at absolute address **8010**, which is the first available word above the user data segment. The **ptz** program is loaded immediately after the last instruction in **main** and hence starts at absolute location **8021**.

The label tables table for the i-code in Figure 7-7 is

```
((main ((main prelude)  .  8010)
        ((main . 0)       .  8012)
        ((main . 1)       .  8016)
```

i-code labels and instructions	Piton instruction	pc
`((main`		
`(dl (main prelude)`	`(prelude)`	
`(cpush_cfp))`		`; 0`
`(move_cfp_csp)`		`; 1`
`(dl (main . 0)`	`(call ptz)`	
`(cpush_*))`		`; 2`
`(pc (main . 1))`		`; 3`
`(jump_*)`		`; 4`
`(pc (ptz prelude))`		`; 5`
`(dl (main . 1)`	`(ret)`	
`(jump_*))`		`; 6`
`(pc (main . 2))`		`; 7`
`(dl (main . 2)`	`(postlude)`	
`(move_csp_cfp))`		`; 8`
`(cpop_cfp)`		`; 9`
`(cpop_pc))`		`;10`
`(ptz`		
`(dl (ptz prelude)`	`(prelude)`	
`(cpush_cfp))`		`; 0`
`(move_cfp_csp)`		`; 1`
`(dl (ptz . 0)`	`(dl loop nil`	
	`(test-nat-and-jump zero`	
	`end))`	
`(tpop{n}_<z>_y))`		`; 2`
`(move_x_*)`		`; 3`
`(pc (ptz . 2))`		`; 4`
`(jump-z_x)`		`; 5`
`(dl (ptz . 1)`	`(jump loop)`	
`(jump_*))`		`; 6`
`(pc (ptz . 0))`		`; 7`
`(dl (ptz . 2)`	`(dl end nil (ret))`	
`(jump_*))`		`; 8`
`(pc (ptz . 3))`		`; 9`
`(dl (ptz . 3)`	`(postlude)`	
`(move_csp_cfp))`		`;10`
`(cpop_cfp)`		`;11`
`(cpop_pc)))`		`;12`

Figure 7-7: Compiler Output for **main** and **ptz**

```
      ((main . 2)      . 8018))
(ptz ((ptz prelude)    . 8021)
      ((ptz . 0)       . 8023)
```

```
((ptz . 1)        . 8027)
((ptz . 2)        . 8029)
((ptz . 3)        . 8031))).
```

For example, the i-code instruction to which the label (**main** . **1**) is attached will be loaded at absolute location **8016**. That is, the absolute location defined by (**main** . **1**) is **8016**. The postlude of **ptz**, at "virtual Piton **pc**" (**ptz** . **3**), starts at **8031**. Inspection of the the i-code reveals that the postlude contains three FM9001 instructions and hence the last instruction in **ptz** is at location **8033**.

Finally, the system data segment is allocated immediately above the program segment. If the maximum control stack size were **5000** and the maximum temporary stack size were **1000**, then the system data link table would be

```
((cstk                  .   8034)
 (tstk                  .  13035)
 (full-ctrl-stk-addr    .  14036)
 (full-temp-stk-addr    .  14037)
 (empty-temp-stk-addr   .  14038)).
```

That is, **cstk** starts immediately above the last FM9001 instruction in **ptz** or at address **8034** and extends for **5000** words. Thus, **tstk** starts immediately above that, at **13035** and extends for **1000** words, etc.

The computation of the four link tables is straightforward. See 'i-link-tables', page 272.

7.6.3. Linking Data Objects

Data objects are linked by the function 'link-data-word' which is defined on page 289. Eight types of data objects are encountered by the linker: the seven Piton data types, **nat**, **int**, **bitv**, **bool**, **addr**, **pc** and **subr**, and the implementation-internal type **sys-addr**. The linker uses the tag word to determine the type of the object and then maps it into a bit vector accordingly. We discuss each in turn.

7.6.3.1. Naturals

Natural numbers are represented in the standard binary notation. Thus, (**nat 3**) is mapped to the 32-bit wide vector

00000000000000000000000000000011.

As previously noted, by this "bit vector" we mean the list of 32 Booleans list (**t, t, f f f** ... **f**).

7.6.3.2. Integers

We use the standard twos-complement representation of integers. Thus, **(int 3)** maps to the same thing as **(nat 3)**, above, and **(int -3)** maps to the bit vector **11111111111111111111111111111101**.

7.6.3.3. Bit Vectors

Piton's bit vector objects are mapped directly to FM9001 bit vectors. Thus,

```
(BITV (0 0 0 0 1 1 1 0 0 0 0 1 1 1 1
       0 0 0 0 0 0 0 0 1 1 1 1 1 1 1 1))
```

is mapped to the bit vector **00001110000111100000000011111111**.

7.6.3.4. Booleans

(bool f) is mapped to the bit vector

00000000000000000000000000000000

and **(bool t)** is mapped to the bit vector

00000000000000000000000000000001.

7.6.3.5. Data Addresses

An object of the form **(addr** (*name* . *n*)**)** is mapped first to a natural number and then to the bit vector representing that natural in binary notation. To map such a data address to a natural number we look in the data link table to get the base address of *name* and add *n* to it. For example, in the link tables shown above, **b** is assigned location **8005**, so **(addr (b . 2))** maps first to **8007** and then to the bit vector **00000000000000000001111101000111**.

7.6.3.6. Program Addresses

An object of the form **(pc** (*name* . *x*)**)** is mapped first to a natural number and then to the corresponding bit vector. The natural number is obtained by finding the label table for *name* in the label tables and then looking up (*name* . *x*) in that table. For example, given the example link tables shown above, **(pc (ptz prelude))** maps to the natural number **8021** and thence to **00000000000000000001111101010101**.

7.6.3.7. Subroutines

An object of the form (**subr** *name*) is mapped first to the natural number associated with *name* in the program link table and then to the corresponding bit vector.

7.6.3.8. System Data Addresses

An object of the form (**sys-addr** (*name* . *n*)) is mapped analogously to data addresses except that the base address of *name* is obtained from the system link table.

7.7. Image Construction

The link-assembly step described above generates two linear lists as its primary output: the contents of the 16 registers and the contents of the initial part of the FM9001 memory. As noted however, the FM9001 requires that the components of the FM9001 state be ram trees. The construction of these trees is the job of the final phase of downloading, called "image construction."

We define ram trees recursively. The concept involves a marking scheme for bit vectors. Recall that at the FM9001 level, a bit vector is a list of ts and fs with "bit 0" or the least significant bit in the 'car'. In this discussion, all bit vectors are of length **32**. Some bit vectors will be marked as 'ram'. In the FM9001 model, that means the bit vector can be both read and written. Other bit vectors may be marked as 'rom', meaning they can only be read. The Piton downloader does not make use of them since they actually reflect the characteristics of the physical memory and cannot be created by "downloading." Formally, both 'ram' and 'rom' are shell constructor functions of one argument each. They construct new shell objects of the same name. If v is a bit vector, then ram (v) is a bit vector marked 'ram' and rom (v) is one marked 'rom'. Since 'ram' and 'rom' are shell constructors the bit vector "inside" can be recovered by applying the accessor for the shell. We sometimes think of these marks simply as colored boxes containing the bit vector.

When we say x is a *ram tree of depth* **0**, we mean x is a 32-bit wide bit vector marked 'ram' or 'rom'. When we say x is a *ram tree of depth* $n + 1$, we mean x is a 'cons' whose 'car' and 'cdr' are each ram trees of depth n.

In Figure 7-8 we show a ram tree of depth **4**. The fringe of the tree is a sequence of bit vectors (each of which is marked 'ram'). We enumerate these tips starting from **0** on the left. The address of each tip can be written in **4**-bit binary since the tree is of depth **4**. Observe that one can obtain the tip at a given **4**-bit address in **4** applications of 'car' and/or 'cdr', using the binary representation as a code for whether to go left ('car', most significant remaining bit **0**) or right ('cdr', bit **1**). For example, r_4 (binary address **0100**) can be reached by starting at the top of the tree in Figure 7-8 and applying 'car' (**0**), 'cdr' (**1**), 'car' (**0**), and then 'car' (**0**).

This model of a sequence of bit vectors is convenient in the formalization of the

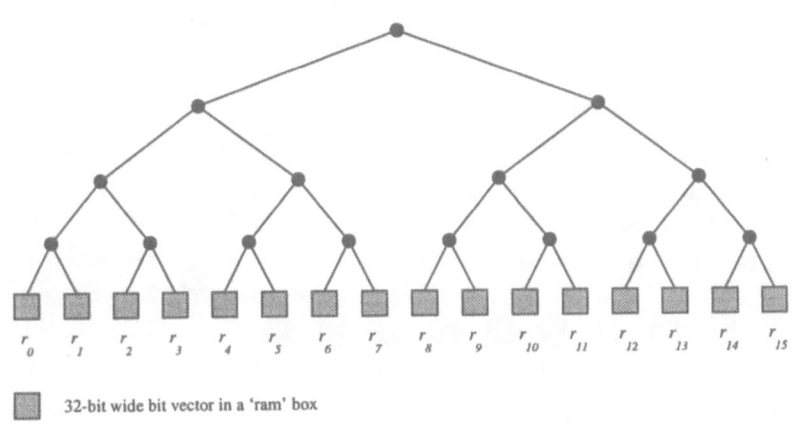

$$r_0 \quad r_1 \quad r_2 \quad r_3 \quad r_4 \quad r_5 \quad r_6 \quad r_7 \quad r_8 \quad r_9 \quad r_{10} \quad r_{11} \quad r_{12} \quad r_{13} \quad r_{14} \quad r_{15}$$

32-bit wide bit vector in a 'ram' box

Figure 7-8: Ram Tree of Depth 4

FM9001 because it allows for faster execution of the formal definition of the FM9001 on concrete data. This is a subtle point. The formal definition of the FM9001 is a specification of a lower level design. That design was ultimately fabricated as the physical device upon which we can run FM9001 machine code programs. But we also run the FM9001 formal model, e.g., to test whether the fabricated device conforms to its design, to simulate machine code programs, and even to perform proof steps involving the application of the 'fm9001' function to constants. When we are running the formal model of the FM9001 (as a recursively defined function in the Nqthm logic) we are concerned about its efficiency as a piece of software. This concern for efficiency motivated the use of ram trees.

The memory of the FM9001 is represented as a ram tree of depth **32**. However, we are generally uninterested in the entire memory tree of the FM9001 and representing it formally as a full binary tree of depth **32** would consume an enormous amount of space. Therefore, the FM9001 memory is represented as a "stubbed out ram tree" of depth **32**. A *stubbed out ram tree of depth n* is a ram tree of depth *n* in which some of the tips and/or interior nodes have been replaced by 'stub' markers. Formally, a 'stub' is another shell constructor of one argument. The argument is a bit vector and in the Piton use, that vector is always all **0**. A stub occurring in a stubbed out ram tree may be thought of as a ram tree of the appropriate depth (so as to complete the containing ram tree) containing the given bit vector (marked 'rom') at every tip.

In Figure 7-9 we show a stubbed out ram tree of depth **4** containing nine explicit bit vectors, m_0-m_8 (each of which is marked 'ram') in memory locations **0** through **8**.

It is now easy to say what the image construction phase of the Piton downloader does with the registers and memory produced by the preceding phases. The list of

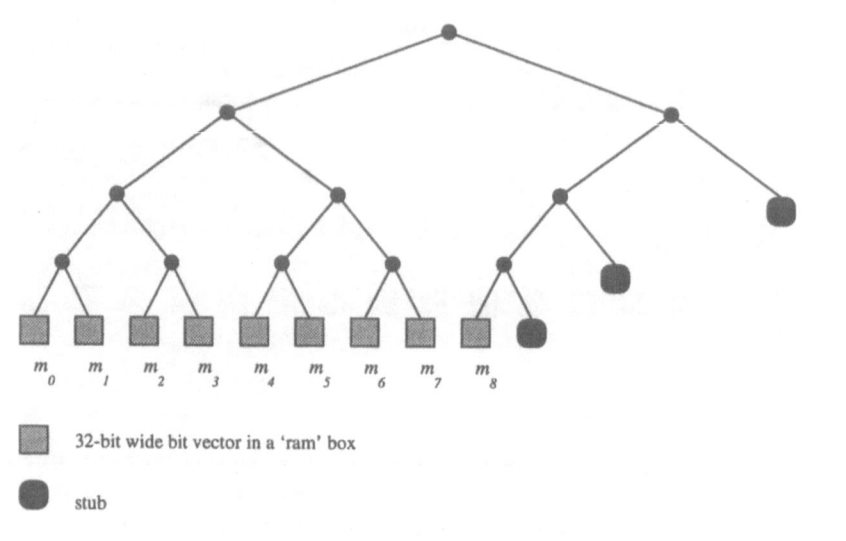

$$m_0 \quad m_1 \quad m_2 \quad m_3 \quad m_4 \quad m_5 \quad m_6 \quad m_7 \quad m_8$$

  32-bit wide bit vector in a 'ram' box

  stub

Figure 7-9: Stubbed Out Ram Tree of Depth 4

registers is converted to a ram tree of depth **4**, with each register marked 'ram'. The memory is converted to a stubbed out ram tree of depth **32**, the first n tips of which are marked 'ram'; the remaining subtrees are stubbed out with the all-**0** vector. Here, n is the sum of the sizes of the boot code, the user data segment, the program segment and the system data segment.

8

Proof of the
Correctness Theorem

In this chapter we discuss how the correctness of our implementation of Piton is proved. We do not give a complete account of the proof but do give the main lemmas and explain how they are put together. The reader interested in obtaining the mechanically checked Nqthm proof script should see page 16.

The FM9001 implementation of Piton is embodied in the function 'load', which is defined as the composition of four functions,

DEFINITION:
load (p, *boot-lst*, *load-addr*)

=

m⇒fm9001 (i⇒m (r⇒i (p⇒r (p)), *boot-lst*, *load-addr*))

one for each of the phases: resource representation ('p⇒r'), compilation ('r⇒i'), link-assembling ('i⇒m'), and ram tree construction ('m⇒fm9001'). The key to our proof of the correctness of the implementation is to prove the correctness of each of the phases separately. But what does it mean to say that the resource representation phase is correct? What does it mean to say that the compiler is correct, in isolation from the link-assembler and FM9001? To formalize the correctness of these phases independently we must formally specify the abstract machines intermediate between Piton and FM9001. That is, we must specify the semantics of the various languages used in the implementation of the Piton compiler and link-assembler. The intermediate machines are called the 'r' or "*resource level*" machine, which interprets r-states, the 'i' or "*instruction (or i-code) level*" machine, which interprets i-states, and the 'm' or "*machine*" level, which interprets m-states.

Once we have formalized these abstract machines we can decompose the proof roughly as shown in Figure 8-1. Observe the four commuting diagrams relating adjacent pairs of abstract machines; each represents a major lemma in the proof. In these diagrams the vertical arrows are all downwardly directed and each corresponds to a phase in 'load'. The composition of the four downward arrows is 'load'. The four small upwardly directed dashed arcs are partial inversions of the corresponding downward arrows. The composition of those arcs represents what we called 'display-fm9001-data-segment' in the main theorem.

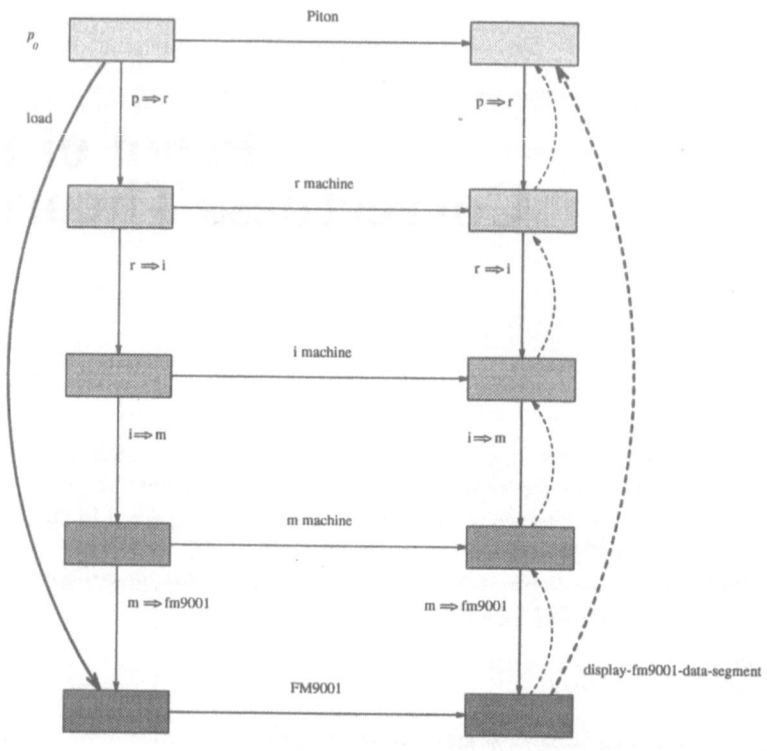

Figure 8-1: A Diagram of the Proof

The decomposition of the proof into "stacking" commuting diagrams is obvious. But recall the question of which way our arrows should be oriented (see page 93). After having gone to the trouble to state the main theorem with an upwardly directed arrow, why not decompose the proof into just four commuting diagrams, each of which has a downwardly directed arrow on the left and an upwardly directed arc on the right? The reason has to do with inductive proof and the fact that the upwardly directed arcs only partially recover the upper state from the lower one.

In Figure 8-2 we exhibit one such diagram (labeled "Theorem") relating two machines, 'p' and 'r' (which here are considered just generic "high" and "low" level machines). Note that the diagram is in the spirit of our main theorem, with its upwardly directed final arc. The arc connecting r_n to p_n means that some portion of the two states are "in correspondence." In our main theorem it means that a certain portion of the FM9001 memory is in correspondence with the data segment of Piton, i.e., that the data segment in p_n can be recovered from that portion of the r_n memory. Now consider trying to prove this theorem by induction on n, the number of computation steps. This is depicted on the right hand side of Figure 8-2. The inductive

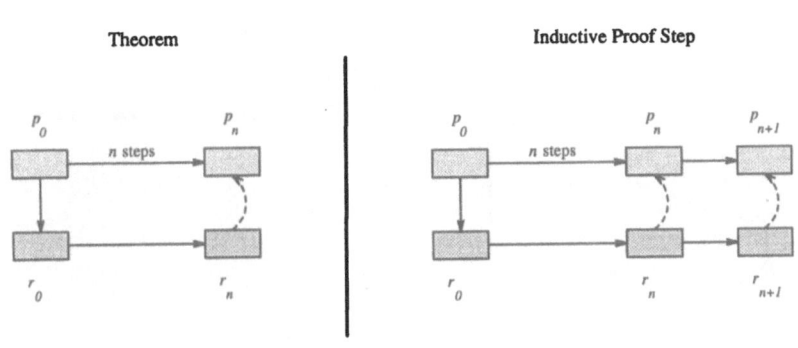

Theorem   Inductive Proof Step

Figure 8-2: An Inductive Proof Attempt

hypothesis tells us that after n steps the two states p_n and r_n are in partial correspondence. We have to show that we can take one more step on each level and arrive at states p_{n+1} and r_{n+1} still in correspondence. But since all we know about p_n and r_n is that the data segment of the former can be recovered from the latter, it is not in general true that the data segments will correspond after one more step. The instruction executed in the $n+1$st step at the p-level may not correspond to that executed in the $n+1$st step at the r-level, since we do not know that the program segments of the two states are in correspondence. In short, the formula indicated by the Theorem diagram in Figure 8-2 is too weak to be proved by induction.

In Figure 8-3 we diagram a different attack on the theorem. The theorem is as above. But this time, the proof is decomposed into two parts, a "one-way correspondence" commuting diagram and a "partial inversion" result. The former establishes that if r_0 is the image of p_0 then r_n is the image of p_n. We call it a "one-way" correspondence because both vertical arrows are downward. The correspondence is "total;" not only do the data segments correspond, but so do the program segments and control segments. The partial inversion lemma establishes that the upwardly directed arrow from the image of p_n accurately recovers the data segment of p_n, i.e., the arc partially inverts the downward arrow. These two lemmas together establish the theorem we wanted. The advantage to this decomposition is that the one-way correspondence theorem can be proved by induction on n.

The induction step is illustrated at the bottom of Figure 8-3. By the induction hypothesis we know that the image of p_n is r_n. We must show that we can take one more step. But since we know the correspondence between p_n and r_n is total, the p-step will "do the same thing" as the r-step. More formally, we can prove the induction conclusion from the induction hypothesis by appealing to a one-way correspondence theorem about the single steppers: p-stepping and then going down produces the same state as going down and then r-stepping. That is the content of the small commuting box linking p_n, p_{n+1}, r_n and r_{n+1}.

The proof of the correctness of Piton on FM9001 is much more complicated than these simple diagrams suggest. For example, one of the component diagrams does

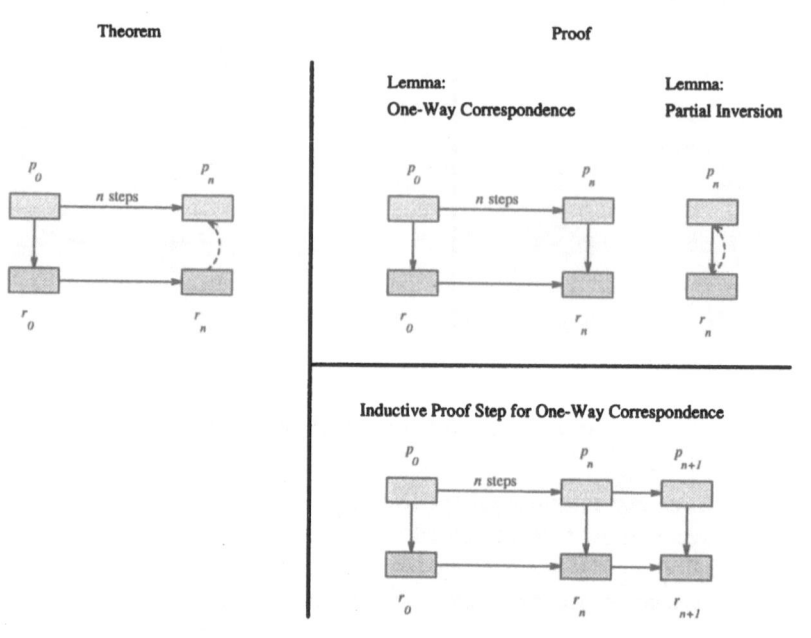

Figure 8-3: An Inductive Proof

not actually commute: the alternative paths do not yield identical states. In another component diagram the number of steps taken by the high-level machine is different than the number taken by the low-level machine. Finally, the theorems corresponding to all of the diagrams have many preconditions and to "stack" the component diagrams we have to prove that the various mappings transform certain properties in ways that allow us to establish these preconditions at each level. But the basic strategy of our proof is made clear by the diagrams and we now describe the semantics of the intermediate machines and state the lemmas formally.

8.1. The R Machine

The 'r' machine is very similar to the 'p' machine in that its programming language is Piton. However, its resources (namely the stacks) are represented in terms of the system data segment and the registers. For example, where the 'p' machine implements a push by 'cons'ing, the 'r' machine does it via decrementing the stack pointer and depositing into the indicated position of the stack array.

Let us consider the Piton **push-constant** instruction. The 'p' machine specification for this instruction is that it increments the program counter and pushes the unabbreviated operand onto the temporary stack. Formally this is rendered

DEFINITION:
p-push-constant-step (*ins*, *p*)
=
p-state (add1-p-pc (*p*),
      p-ctrl-stk (*p*),
      push (unabbreviate-constant (cadr (*ins*), *p*), p-temp-stk (*p*)),
      p-prog-segment (*p*),
      p-data-segment (*p*),
      p-max-ctrl-stk-size (*p*),
      p-max-temp-stk-size (*p*),
      p-word-size (*p*),
      ' run)

The function 'push' here is just 'cons'; the temporary stack is a list and the new object is consed onto the front of that list.

    The 'r' machine specification of **push-constant** is

DEFINITION:
r-push-constant-step (*ins*, *r*)
=
r-state (add1-r-pc (*r*),
      r-cfp (*r*),
      r-csp (*r*),
      push-stk (r-tsp (*r*)),
      r-x (*r*),
      r-y (*r*),
      r-c-flg (*r*),
      r-v-flg (*r*),
      r-n-flg (*r*),
      r-z-flg (*r*),
      r-prog-segment (*r*),
      r-usr-data-segment (*r*),
      deposit (**if** cadr (*ins*) = **' pc then** add1-r-pc (*r*)
             elseif nlistp (cadr (*ins*))
             then pc (cadr (*ins*), r-current-program (*r*))
             else cadr (*ins*) **endif**,
           push-stk (r-tsp (*r*)),
           r-sys-data-segment (*r*)),
      r-word-size (*r*),
      ' run)

    The function 'push-stk' used above decrements a tagged address pair by one, e.g., push-stk (**' (sys-addr (tstk . 25)))** is equal to **' (sys-addr (tstk . 24))**. Deposit (*obj*, *addr*, *segment*) writes *obj* into address *addr* of *segment*, where *addr* is a tagged address pair (specifying a name and an offset) and *segment* is an association list pairing area names with arrays.

    Observe that the 'r' machine specification of **push-constant** increments the program counter, decrements the 'r-tsp' register, and writes the unabbreviated operand to the system data segment location addressed by the decremented 'r-tsp' register. That is, the 'r' machine implements the Piton instructions using the resources of FM9001.

The situation is a little more subtle than this example might suggest. Consider the Piton **push-local** instruction. The 'p' machine specification is

DEFINITION:
p-push-local-step (*ins*, *p*)
=
p-state (add1-p-pc (*p*),
        p-ctrl-stk (*p*),
        push (local-var-value (cadr (*ins*), p-ctrl-stk (*p*)), p-temp-stk (*p*)),
        p-prog-segment (*p*),
        p-data-segment (*p*),
        p-max-ctrl-stk-size (*p*),
        p-max-temp-stk-size (*p*),
        p-word-size (*p*),
        'run)

The 'r' machine specification is

DEFINITION:
r-push-local-step (*ins*, *r*)
=
r-state (add1-r-pc (*r*),   ; (a)
        r-cfp (*r*),
        r-csp (*r*),
        push-stk (r-tsp (*r*)),   ; (b)
        add-addr (r-csp (*r*),   ; (c)
                offset-from-csp (cadr (*ins*),
                                r-current-program (*r*))),
        r-y (*r*),
        r-c-flg (*r*),
        r-v-flg (*r*),
        r-n-flg (*r*),
        r-z-flg (*r*),
        r-prog-segment (*r*),
        r-usr-data-segment (*r*),
        deposit (fetch (add-addr (r-csp (*r*),   ; (d)
                            offset-from-csp (cadr (*ins*),
                                            r-current-program (*r*))),
                    r-sys-data-segment (*r*)),
            push-stk (r-tsp (*r*)),
            r-sys-data-segment (*r*)),
        r-word-size (*r*),
        'run)

The 'add-addr' expression, which occurs twice above (in lines (c) and (d)), computes the address at which the value of the indicated local variable is stored. The address is computed by adding to the current control stack pointer—the contents of the 'r-csp' register—the offset of the local variable among the current program's local variables. Call that address *addr*.

An informal reading of the 'r' machine specification for **push-local** is that it (a) increments the program counter, (b) decrements the 'r-tsp' register, (c) sets the 'r-x' register to *addr* and (d) deposits into the system data area at the new top of the temporary stack the contents of *addr*. Steps (a), (b), and (d) are intuitively necessary and sufficient.

Why, however, does the 'r' machine set the 'r-x' register? It turns out that no 'r' machine instruction inspects the values of the temporary registers, 'r-x', 'r-y' and the four flags. However, many of the instructions set those registers and flags. Why? The answer is that the 'r' machine does more than just implement the Piton instructions with the FM9001 resources. It implements the Piton instructions with the FM9001 resources *in exactly the same way our compiler does*. We explain why later (see page 149).

For example, the compiled code for **push-local** is generated by

DEFINITION:
icode-push-local (*ins*, *pcn*, *program*)
=
list(' (move_x_*),
    tag('nat, offset-from-csp (cadr (*ins*), *program*)),
    ' (add_x{n}_csp),
    ' (tpush_<x{s}>))

Observe that this code moves into the **x** register the offset of the required local, then adds the **csp** register to that, leaving the result in **x**, and then pushes the contents of the address in **x** onto the temporary stack. Thus, the implementation of **push-local** has the additional side-effect on FM9001 of setting the **x** register to the address of the local pushed. The 'r' machine specification of **push-local** faithfully describes this implementation of **push-local**.

Formally, r(*r*, *n*) is defined to step forward *n* times from the r-state *r*, using the function 'r-step'. 'R-step' checks the precondition of the current instruction and, if it is satisfied, produces the next state with the step function for the current instruction.

The 'r' machine can thus be imagined by changing the 'p' machine as follows. First, replace all references to the abstract stacks by fetches and deposits on the registers and system data segment so as to mimic at the r-level what the *p* machine does. Second, add the six temporary registers and flags so that the 'r' machine correctly describes the final values of those FM9001 resources at the conclusion of the i-code generated for each instruction.

The reader familiar with proofs about recursive functions will immediately recognize one important role of the 'r' machine. It allows us to factor the proof into two parts: prove that 'p' "is" 'r' and then prove that 'r' "is" FM9001. In this view, 'r' appears merely to be a convenient way station for the weary traveler. Given enough patience and energy, the proof appears possible without the introduction of 'r'. But this is not so. 'R' or something very much like it is a necessary stop on the way from 'p' to FM9001. We explain this after introducing the next machine.

8.2. The I Machine

The 'i' machine operates on i-states. Recall that i-states are like r-states except that the programs are written in i-code, not Piton. Thus, the 'i' machine interprets i-code using the same stack representation as the 'r' machine. The 'i' machine is defined in a style very similar to our other interpreters. By definition, i(i, n) steps the i-state i forward n times, using the precondition and step functions specific to the current instruction. We illustrate the 'i' machine by simply exhibiting the step functions for several of the 87 i-code instructions.

Here is the step function for the i-code instruction **move_x_***, which moves the next word of the instruction stream into register **x** and increments the program counter by two:

DEFINITION:
i-move_x_*-step (i)
=
i-state (add2-i-pc (i),   ; increment **pc** by 2
        i-cfp (i),   ; **cfp** unchanged
        i-csp (i),   ; **csp** unchanged
        i-tsp (i),   ; **tsp** unchanged
        i-nextword (i),   ; move next word into **x**
        i-y (i),   ; **y** unchanged
        i-c-flg (i),   ; flags unchanged
        i-v-flg (i),
        i-n-flg (i),
        i-z-flg (i),
        i-prog-segment (i),   ; program and data unchanged
        i-usr-data-segment (i),
        i-sys-data-segment (i),
        i-word-size (i),
        'run)

The function 'i-nextword', used above, fetches the contents of the program segment address one greater than the current program counter,

DEFINITION:
i-nextword (i) = unlabel (fetch (add1-i-pc (i), i-prog-segment (i)))

'I-move_x_*-step' deposits the object fetched into the **x** register and increments the program counter by two.

Below we exhibit the semantics for the i-code instruction **add_<tsp>{n}_x{n}**. Informally, it replaces the top of the temporary stack by the natural number sum of the current top and the contents of the **x** register, both of which must be naturals. The precondition for the instruction checks that both operands of the addition are tagged naturals and that their sum is representable.

DEFINITION:
i-add_<tsp>{n}_x{n}-step (*i*)
=
i-state (add1-i-pc (*i*),
    i-cfp (*i*),
    i-csp (*i*),
    i-tsp (*i*),
    i-x (*i*),
    i-y (*i*),
    i-c-flg (*i*),
    i-v-flg (*i*),
    i-n-flg (*i*),
    i-z-flg (*i*),
    i-prog-segment (*i*),
    i-usr-data-segment (*i*),
    deposit (tag (**'nat**,
                untag (fetch (i-tsp (*i*), i-sys-data-segment (*i*)))
           +   untag (i-x (*i*))),
        i-tsp (*i*),
        i-sys-data-segment (*i*)),
    i-word-size (*i*),
    'run)

Observe that the new system data segment in the i-state above is obtained by writing (with 'deposit') the tagged sum to the old system data segment at the address in the **tsp** register.

Recall that i-code instructions are mapped into machine code via the table shown on page 119. Some distinct i-code instructions are mapped to the same assembly instruction (and hence to the same machine instruction). For example, both **add_<tsp>{n}_x{n}**, shown above, and its integer counterpart, **add_<tsp>{i}_x{i}**, are both mapped to **(add () (tsp) x)**. The 'i' machine differentiates these two i-code instructions. In particular, the 'i' machine semantics for the integer version of the instruction uses integer addition, 'iplus', and tags the result with **int** rather than **nat**. Similarly, the i-code instruction **move_x_x** and the i-code instruction **int-to-nat** map to **(move nil x x)**. The former instruction is a no-op on the 'i' machine. The latter instruction changes the top of the temporary stack from from a tagged **int** to a tagged **nat**.

The main point is that the 'i' machine provides abstract data types, even though its instruction set is in 1:1 correspondence with FM9001. The compiler will be proven correct with respect to the 'i' machine semantics, not the FM9001 semantics (because we will not involve the linker in the correctness of the compiler). It is not until we drop down through the linker that addition on the naturals becomes the same operation as addition on the integers and the conversion of some integers to naturals becomes a no-op (through the wonders of twos complement representation). Given that the two operations are different at the i-level, they must have different names. The "mnemonic device" of annotating otherwise identical i-code instructions with data type information such as **{n}** and **{i}** is fundamental to our strategy of separating the compiler proof from the link-assembly proof.

This is an important point that bears repeating because it impacts the way compilers should be written. Suppose the compiler had been defined so as to generate assembly language directly rather than the annotated assembly language of i-code. For example, suppose the compiler generated (**move nil x x**) both where it now generates **move_x_x** and where it generates **int-to-nat**. This is certainly adequate if we simply intend to link-assemble the output of the compiler and run it. But if we adopt this approach, then we must prove the compiler correct in conjunction with the linker, which means we must prove a theorem directly relating the compiler's input to the linker's output, i.e., relating Piton programs and data to their binary compiled images. This vastly complicates the compiler proof. The complication does not come solely from the introduction of the bit-vector representation of integers and naturals but also from the bit-vector representation of addresses, including program counters. These addresses are readily related to the (now intermediate) i-code but difficult to relate to the Piton source code.

Why can we not separate the compiler proof from the linker proof if the compiler generates unannotated assembly code? To separate the two, we must give semantics to the unannotated assembly code. What semantics do we give to (**move nil x x**) in the case where the top of the stack is a tagged **int**? Do we make that instruction a no-op or do we make it retag the top of the stack? Without the data type annotations we cannot give semantics to the assembly code in isolation from the linker.

One 'i' machine instruction deserves special mention, namely **xor_<z>_<tsp>_x**, which assembles to (**xor (z) (tsp) x**). This instruction pushes onto the top of the temporary stack the exclusive-or of two arbitrary objects and sets the **z** flag to indicate whether the result is the **0**-vector. At the machine code level, exclusive-or is the fundamental test of identity; the exclusive-or of two bit vectors is the **0**-vector if and only if the two bit vectors are identical. The **xor_<z>_<tsp>_x** instruction is used in the implementation of the Piton **eq** instruction, which determines if two arbitrary Piton objects are equal.

The question is, on the 'i' machine, what should **xor_<z>_<tsp>_x** push onto the temporary stack if the two objects tested are not identical? That is, what is the semantics of **xor_<z>_<tsp>_x**, particularly in the case when two distinct objects are compared? Recall that i-level objects are symbolic entities, e.g., (**addr (delta1 . 25)**) and (**addr (finstr . 17)**). What "should" **xor_<z>_<tsp>_x** do with these two? It is clear that their exclusive-or is non-0 and hence the **z** flag should be cleared. But what value should the 'i' machine push as the result of the exclusive-or? From the perspective of proving that **eq** is correctly implemented on the 'i' machine, it does not matter what exclusive-or does, as long as the **z** flag is properly set, because in the code sequence in which **xor_<z>_<tsp>_x** is used, namely that generated by 'icode-eq' (see page 275), the result of the exclusive-or is discarded. But whatever semantics we choose to give **xor_<z>_<tsp>_x** we must insure that it is correctly implemented by the corresponding FM9001 instruction.

We decided to define **xor_<z>_<tsp>_x** to link-assemble the two operands, i.e., to convert them to bit vectors via the same process used by the link-assembler, to push the bit-vector result of that exclusive-or, and to set the **z** flag according to whether the two operands were identical. This makes it easy to prove that `xor_` `<z>_<tsp>_x` is correctly implemented (after linking) by `(xor (z) (tsp)` `x)`. But it forces the 'i' machine to have the *load-addr* as a parameter, because the bit-vector representations of the operands cannot be determined without knowing where the image is to be placed.

In some ways the 'i' machine is similar to FM9001. The main similarity comes from the fact that each i-code instruction translates into a single FM9001 instruction. Unlike FM9001, i-code instructions are symbolically represented (rather than encoded as bit vectors) and all data, addresses, and program counters are tagged and represented symbolically (rather than as bit vectors). Unlike FM9001, i-code instructions check for violations of type and resource preconditions and produce erroneous states when violations are found. This insures that program space is not overwritten and that no operation is done that exposes the concrete representation of the data objects.

8.3. The M Machine

The next phase in downloading is link-assembling. As suggested by the name, the link-assembler assigns absolute locations to all objects and resources and maps from symbolic instructions to binary machine code. At the conclusion of link-assembling the distinctive entities of the i-level—instructions, programs, system data, labels, abstract data types, etc.— reside in an undistinguished linear sequence of 1s and 0s which can be arbitrarily read, written, or executed. The machine that operates on this binary state can be made to violate our invariants. For example, the machine allows programs to overwrite themselves or to clear register 2 (our control stack pointer). Because the link-assembler allocates the symbolic resources to different physical locations but allocates linearly accessed ones (such as programs, stacks, and arrays) contiguously, and because it replaces symbolic addresses by accurate absolute ones, we will be able to prove that such violations do not occur. More generally, we prove that there is a step-by-step correspondence between the 'i' machine and the lower level machine.

Intuitively, that lower level machine is the FM9001. But recall that the FM9001 uses ram trees to represent the register file and memory. This is a needless complication and we sweep it aside by providing the 'm' machine. The 'm' machine is the FM9001 except that it uses linear lists to represent the registers and memory.

8.4. The One-Way Correspondence Lemmas

We now have five machines, 'p', 'r', 'i', 'm' and 'fm9001'. Between each adjacent pair of machines we have a one-way correspondence lemma.

8.4.1. From P to R

Our description of the 'r' machine makes it intuitively clear that it is, in some informal sense, equivalent to the 'p' machine, under a certain mapping between the two different ways the stacks are represented. That mapping, or at least the half of it that goes from the abstract representation used by 'p' to the more concrete representation used by 'r', is in fact just 'p⇒r'.

Consider two alternative r-states. The first, which we shall call r_n, is obtained by running the 'p' machine n steps from some p-state p_0 and then mapping down with 'p⇒r'. The second, which we shall call r_n', is obtained by mapping p_0 down with 'p⇒r' and then running the 'r' machine n steps. See Figure 8-4.

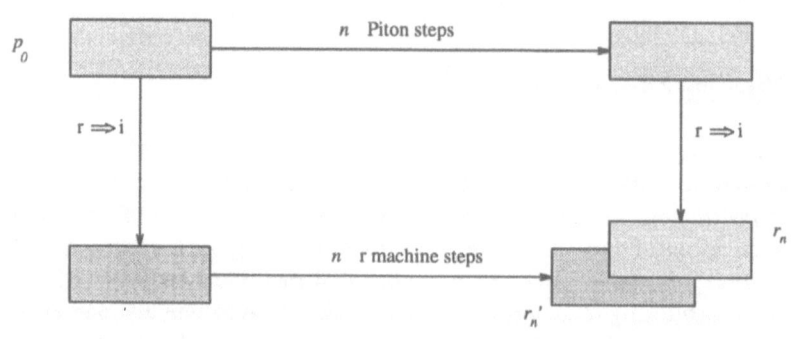

Figure 8-4: P Down to R

What is the relation between r_n and r_n'?

They are not always equal. Consider the **x** register. In r_n the **x** register is set to the natural number 0 because that is how 'p⇒r' initializes it. But in r_n', the value of the **x** register is determined by the last instruction executed which set that temporary register.

Are r_n and r_n' equal if we ignore the six temporary registers and flags? No. Consider the temporary stack area, in particular, that region of the area beyond the current **tsp** register—the "inactive" region "above" the top-of-stack. In r_n that region consists entirely of 0's, because that is how 'p⇒r' initializes it. In r_n' that region contains whatever data was put there by the instructions executed by the 'r' machine. How can data be put above the top of the stack? On the 'r' machine that is possible, by pushing the data onto the temporary stack and then popping the stack.

Are r_n and r_n' equal if we ignore the six temporary registers and flags and the two stack regions beyond their respective stack pointers? Yes. We define the predicate 'r-equal' to be this sense of weak equality. We say that the **x** and **y** registers, the condition code registers, and the inactive stack regions of an r-state are *hidden resources* of the 'r' machine. We say that the rest of the r-state is *visible*. Two r-states are 'r-equal' if and only if their visible resources are identical.

The Piton machine can be related to the 'r' machine via 'p⇒r' with the following one-way correspondence theorem.

THEOREM 1:

$$(\quad \text{proper-p-statep} (p)$$
$$\wedge \quad \text{p-loadablep} (p, \textit{load-addr})$$
$$\wedge \quad (\neg \, \text{errorp} (\text{p-psw} (\text{p} (p, n)))))$$
$$\rightarrow \text{r-equal} (\text{p⇒r} (\text{p} (p, n)), \text{r} (\text{p⇒r} (p), n))$$

This theorem may be read as follows. Suppose p is a proper p-state that is loadable and non-erroneous at Piton step n. Let r_n and r_n' be as above. Then r_n is 'r-equal' to r_n'.

The proof of the one-way correspondence theorem from 'p' to 'r' is by induction on n. The induction step requires two key lemmas. The first is that the one-way correspondence holds for the single steppers of the two machines,

THEOREM:

$$(\text{proper-p-statep} (p) \wedge (\neg \, \text{errorp} (\text{p-psw} (\text{p-step} (p)))))$$
$$\rightarrow \text{r-equal} (\text{p⇒r} (\text{p-step} (p)), \text{r-step} (\text{p⇒r} (p)))$$

The second is that 'r-equal' is a congruence relation for 'r',

THEOREM:

$$(\quad \text{proper-r-statep} (r_1, \textit{load-addr})$$
$$\wedge \quad \text{proper-r-statep} (r_2, \textit{load-addr})$$
$$\wedge \quad \text{r-equal} (r_2, r_1))$$
$$\rightarrow \text{r-equal} (\text{r} (r_2, n), \text{r} (r_1, n))$$

i.e., if two proper r-states are 'r-equal' then so are the states produced by running each with 'r'. We have not discussed proper r-states before. Suffice it to say that they are to r-states what loadable proper p-states are to p-states. That is, the proper r-states are those loadable r-states in which all the components are well-formed and compatible, where all the checks are made on the r-level representation of the stacks instead of the p-level.

Other lemmas used in the proof of the one-way correspondence theorem include

- If a p-state is proper and can be stepped non-erroneously then the result is proper.

- If a proper p-state is loadable and can be stepped non-erroneously then the result is loadable.

- If a p-state is erroneous then running it with the 'p' machine is erroneous.

- If a p-state is proper and loadable then its image under 'p⇒r' is proper (at the r-level).

- If an r-state is proper (at the r-level) then stepping it is proper (at the r-level).

- 'r-equal' is reflexive and transitive.

We leave the proof of the one-way correspondence theorem to the reader. We urge the reader to construct it.

We briefly discuss the proof of the two key lemmas noted above.

The proof of the first lemma, the one-way correspondence "step," is by case analysis on the current instruction of p. For each of the Piton instructions we prove the corresponding one-way correspondence lemma. Here, for example, is the one-way correspondence lemma for the **push-constant** instruction.

THEOREM:

$($ (p-psw (p)) = '**run**)
$\wedge$ (car (p-current-instruction (p))) = '**push-constant**)
$\wedge$ proper-p-statep (p)
$\wedge$ p-push-constant-okp (p-current-instruction (p), p))
$\rightarrow$ r-equal (p$\Rightarrow$r (p-push-constant-step (p-current-instruction (p), p)),
       r-push-constant-step (p-current-instruction (p), p$\Rightarrow$r (p))))

These instruction-level lemmas are really the heart of the proof of the one-way correspondence theorem. To prove these lemmas we had to prove the basic facts relating the abstract representation of stacks to the concrete one. For example, we had to prove such facts as

- Pushing an object onto a stack and then mapping that stack down to the r-level produces the same thing as mapping the original stack down to the r-level and then decrementing the stack pointer address and writing the object at the indicated location.

- Popping an object from a stack and then mapping that stack down produces the same thing as mapping the original stack down and then incrementing the stack pointer.

- Building a new frame as specified by the **call** instruction, pushing it onto the control stack and then mapping that stack down produces the same thing as mapping the stack down and then pushing a certain sequence of objects.

- Finding the value of a local variable by looking in the bindings field of the top-frame of the control stack produces the same object as mapping the control stack down to the r-level and fetching the object at a certain offset from **csp**.

Of course, the first three "facts" are not actually valid unless by the phrase "produces the same thing as" we mean "produces the same active stack region as."

The proof of the second key lemma, that 'r-equal' is a congruence relation for 'r', was broken down first into the analogous fact that 'r-equal' is a congruence relation for 'r-step' and then into a case for each Piton instruction.

It is interesting to note that for the 'p$\Rightarrow$r' proofs it is not important what the r-level machine does with the hidden resources. It is only important that the values of the hidden resources never affect the values of the visible ones.

Before leaving Theorem 1 we note an important corollary.

COROLLARY 1:

 (proper-p-statep (p)

 $\wedge$ p-loadablep $(p, load\text{-}addr)$

 $\wedge$ $(\neg$ errorp (p-psw (p $(p, n)))))$

$\rightarrow$ (r-usr-data-segment (p$\Rightarrow$r (p $(p, n)))$ = r-usr-data-segment (r (p$\Rightarrow$r $(p), n)))$

This follows immediately from Theorem 1 since the 'r-usr-data-segment's of two 'r-equal' states are identical (because those segments contain no hidden resources).

8.4.2. *From R to I*

We now move down one step and consider the relation between the 'r' machine and the 'i' machine. This can be formalized with 'r$\Rightarrow$i'. Recall that the difference between r-states and i-states is that the former contain Piton programs and the latter contain i-code programs. 'R$\Rightarrow$i' is the compiler. The appropriate one-way correspondence theorem is shown below.

THEOREM 2:

 (proper-r-statep $(r, load\text{-}addr)$

 $\wedge$ $(load\text{-}addr \in \mathbf{N})$

 $\wedge$ $(\neg$ errorp (r-psw (r $(r, n)))))$

$\rightarrow$ (r$\Rightarrow$i (r $(r, n))$ = i (r$\Rightarrow$i (r), clock $(r, n), load\text{-}addr))$

This theorem may be read as follows: Suppose r is a proper r-state and that the result of running n steps at the r-level starting at r is non-erroneous. Then consider two alternative i-states. The first is obtained by running r at the r-level n steps and mapping down with 'r$\Rightarrow$i'. The second is obtained by mapping r down to an i-state and then running it clock (r, n) steps at the i-level. These two i-states are actually equal.

The 'clock' function is just a modification of the 'r' machine that counts the number of i-code instructions executed on behalf of each Piton instruction.[14]

The one-way correspondence result for 'r$\Rightarrow$i' is beautiful because it is a strict equality, not a weak equality as in the 'p$\Rightarrow$r' case. This is so because the 'r' machine accurately accounts for the use of all the resources of FM9001—even the hidden ones.

Another interesting aspect of this result is that the number of steps taken by the upper level machine, 'r', is in general different than that taken by the lower level machine, 'i'. In all our other one-way correspondence theorems, the upper and lower machines take exactly the same number of steps. Of course, this is unavoidable since it is at this level that we change programming languages.

The proof is by induction on n again and the key lemma is the one-way correspondence theorem for 'r-step',

---

[14]The previously discussed expression fm9001-clock (p, n) is in fact defined to be clock (p$\Rightarrow$r $(p), n)$. The remarks made earlier suggesting that 'fm9001-clock' was derived from 'p' were true in the sense that 'clock' is derived from 'r' and 'r' is derived from 'p'. We could not say this earlier because the 'r' machine had not been introduced.

THEOREM:

    (proper-r-statep $(r, load\text{-}addr)$

    $\wedge$ $(load\text{-}addr \in N)$

    $\wedge$ $(\neg$ errorp (r-psw (r-step (r))))))

$\rightarrow$ (r$\Rightarrow$i (r-step (r))) = i (r$\Rightarrow$i (r), r-step-clock (r), $load\text{-}addr$))

Note that on the left-hand side of the conclusion we have a single 'r-step', while on the right-hand side we run 'i' a certain number of steps, namely r-step-clock (r).

To use this 'r-step' lemma to prove the one-way correspondence theorem we need the observation

THEOREM:

i $(i, x + y, load\text{-}addr)$ = i (i $(i, x, load\text{-}addr)$, y, $load\text{-}addr)$

which lets us piece together many short runs of 'i'—each over the block of compiled code corresponding to one Piton instruction—into a single long run of 'i'. This lemma is easy to prove and we do not discuss it further.

We turn instead to the proof of the one-way correspondence theorem for 'r-step'. That theorem is the heart of the compiler proof. Once again, we split the theorem into a separate case for each Piton instruction. Here is the case for **push-local**.

THEOREM:

    ((r-psw (r) = **'run**)

    $\wedge$ (car (r-current-instruction (r))) = **'push-local**)

    $\wedge$ proper-r-statep $(r, load\text{-}addr)$

    $\wedge$ $(load\text{-}addr \in N)$

    $\wedge$ r-push-local-okp (r-current-instruction (r), r))

$\rightarrow$ (r$\Rightarrow$i (r-push-local-step (r-current-instruction (r), r))

    =   i (r$\Rightarrow$i (r),

            r-push-local-step-clock (r-current-instruction (r), r),

            $load\text{-}addr$))

Observe that the hypotheses include the assumption that the opcode of the current instruction at the r-level is **push-local**. The conclusion involves running the 'i' machine for a certain number of clock ticks on the i-state r$\Rightarrow$i (r). The function 'r-push-local-step-clock' determines the number of i-code instructions executed for this particular **push-local** execution. In the case of **push-local** the number of i-code instructions is 3 (since there are no branches in the i-code generated for **push-local** the number of clock ticks for that instruction is constant). Thus, the right-hand side of the conclusion of the above theorem is equivalent to i (r$\Rightarrow$i (r), **3**).

The proof of the theorem requires the ability to do two kinds of reasoning. First, given a particular Piton instruction as the current instruction at the r-level, we must be able to deduce which i-code instructions will be executed when we run the i-level image of the r-state. Second, given the current i-level instruction, we must be able to step an i-state symbolically. The second kind of reasoning is not hard—it is just expanding the definition of 'i'—provided we can deduce what the current i-level instruction of an i-state is. The initial i-state in which we are interested is always the image of an r-state. But after we have executed the first i-code instruction in that state, we generally have an i-state that is not the image of any r-state. We must be able to deduce what the next i-code instruction is from the fact that (a) we started with the image of an r-state, (b) we knew the opcode of the current instruction of that r-state, and (c) we are executing only as many instructions as generated for that opcode.

For example, from the assumption that the current instruction at the r-level is a **push-local** we must be able to show that the next three i-code instructions to be executed will be **(move_x_*)**, **(add_x{n}_csp)**, and **(tpush_<x{s}>)**.

The generalized version of this is that in the image of an r-state the program counter points to the beginning of the block of i-code instructions generated for the current r-level instruction. We formalize this as

THEOREM:
let *icode* **be** icode (get (offset (r-pc (r)),
                            program-body (definition (adp-name (untag (r-pc (r))),
                                                              r-prog-segment (r)))),
                   offset (r-pc (r)),
                   definition (adp-name (untag (r-pc (r))), r-prog-segment (r))),
    i-pc **be** untag (r⇒i_pc (r-pc (r), r-prog-segment (r))),
    i-prog-segment **be** icompile (r-prog-segment (r))
in
    (proper-r-statep (r, *load-addr*) ∧ (n < length (*icode*)))
→ (fetch-adp (add-adp (*i-pc*, n), *i-prog-segment*) = get (n, *icode*)) **endlet**

This theorem, while complicated, expresses a beautiful fact. As stated above, the theorem introduces three variables to stand for more complicated expressions. The variable *icode* may be informally read as "the i-code generated for the current instruction of the r-state r." The variable *i-pc* may be read as "the address pair pointing to the current instruction in the image of r under 'r⇒i'." The variable *i-prog-segment* may be read as "the program segment of the image of r under 'r⇒i'." Thus, the body of the theorem above

THEOREM:
    (proper-r-statep (r, *load-addr*) ∧ (n < length (*icode*)))
→ (fetch-adp (add-adp (*i-pc*, n), *i-prog-segment*) = get (n, *icode*))

may be read as follows: Let r be a proper r-state. Suppose n is a natural number less than the length of the i-code generated for the current instruction, *ins*, of r. Let i be the image of r under 'r⇒i'. Increment the program counter of i by n and fetch from the program segment of i. What do you get? The nth instruction in the i-code generated for *ins*.

This theorem lets us determine the i-code instruction to be executed in any i-state produced by stepping forward from the image of an r-state, provided one has not stepped so far forward that the program counter has been pushed beyond the basic block of code generated for the current r-level instruction. The theorem is the key to our 'r⇒i' level proofs.

Our proof also relies on such lemmas as

- The fundamental properties of fetch and deposit, e.g., that fetch $(a_1,$ deposit $(val, a_2, segment))$ is *val* if a_1 is a_2 and is fetch $(a_1, segment)$ otherwise. These lemmas enable the i-code to create a complicated state by performing several simpler transformations sequentially.

- If x is a legal r-level object in some r-state then it is also a legal i-level object in the image of that r-state. This theorem lets us establish the i-level preconditions from the r-level preconditions.

In addition, there were many facts that were needed to handle the compiled code for specific instructions.

For example, the i-code for the **test-int-and-jump** instruction in the case where the test is **pos** is generated by

```
list('(tpop{i}_<zn>_y),                                      ; 1
     '(move_x_*),                                             ; 2
     tag('pc, cons(name(program), 1+pcn)),                   ; 3
     '(jump-n_x),                                             ; 4
     '(jump-z_x),                                             ; 5
     '(jump_*),                                               ; 6
     pc(caddr(ins), program))                                 ; 7
```

This code may be read as follows: (1) Pop the stack into **y** and set both the zero and negative flags. (2-3) Move into **x** the address of the beginning of the next Piton instruction. (4) Jump to the address in **x** if the negative flag is on. (5) Jump to the address in **x** if the zero flag is on. (6-7) Jump unconditionally to the label in the **test-int-and-jump** instruction. The claim is that this code jumps to the indicated label if and only if the top of the stack is positive. The proof of correctness requires the (trivial) fact that an integer is positive if and only if it is neither zero nor negative.

More interesting is the i-code for the **lt-int** instruction. It is supposed to determine if the top two elements of the stack are in the "less than" relation (where the deeper element is the lesser). This instruction is supposed to work correctly for any two representable integers. The i-code is

```
'((tpop_x)                                                   ; 1
  (sub_<nv>_<tsp>{i}_x{i})                                   ; 2
  (move_<tsp>_*)                                             ; 3
  (bool f)                                                   ; 4
  (move-v_<tsp>_*)                                           ; 5
  (bool t)                                                   ; 6
  (move_x_*)                                                 ; 7
  (bool f)                                                   ; 8
  (move-n_x_*)                                               ; 9
  (bool t)                                                   ;10
  (xor_<tsp>{b}_x{b})).                                      ;11
```

This code (1) pops the stack into **x** and then (2) subtracts **x** from the new top of the stack, storing the result on top of the stack and setting both the negative and the overflow flags. Instruction (3-4) writes **f** to the top of the stack and instruction (5-6) overwrites it with **t** if the overflow flag is on. Instruction (7-8) puts an **f** in **x**, which is overwritten with **t** by instruction (9-10) if the negative flag is set. Finally, at (11) the code computes the exclusive-or of the top of the stack and **x**, leaving the result on the top of the stack. The claim is: the stack has been popped twice and a **t** or an **f** has been pushed according to whether the two popped elements were in the required less than relation. This is a non-trivial claim and depends upon the fact that after computing $i - j$ in the twos-complement representation, the exclusive-or the negative and overflow flags is equivalent to $i < j$.

The hardest instruction to handle was, of course, **call**, where an arbitrary number of i-code instructions must be executed. The proof of the one-way correspondence lemma for **call** required two inductively proved lemmas, one to show that the execution of the code generated by 'generate-prelude1' correctly pushes the initial values of the temporaries and the other to show that the execution of the code generated by 'generate-prelude2' correctly pushes the actual values of the formals and removes them from the temporary stack.

We conclude this discussion of the 'i' machine by reconsidering the role of the 'r' machine. Is the 'r' machine just a convenient way station between 'p' and FM9001? No. The essential contribution of 'r' is that it separates the hidden resources from the visible ones and it establishes that the hidden ones can be used arbitrarily within the "basic block" of a Piton instruction at the i-level. We elaborate this point below.

Consider the attempt to go directly from the p-level to the i-level via what we shall call 'p⇒i'. See Figure 8-5. It is not the case that i_k is the same as i_k', because of

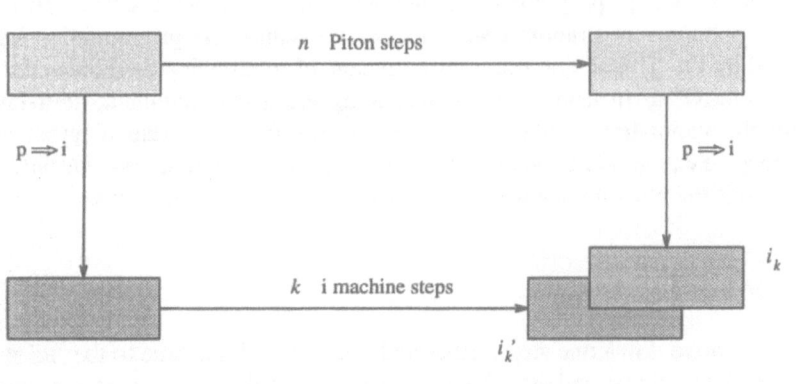

Figure 8-5: P Down to I without R

hidden resources. So we tackle the problem as we did before, by introducing the notion of "i-equivalence," which checks that the visible part of two i-states are equal. The one-way correspondence formula relating 'p' to 'i' expresses the i-equivalence of i_k and i_k'. The proof requires induction on the number, n, of Piton instructions executed. The induction hypothesis establishes that the i-states are i-equivalent after $n - 1$ Piton steps. We need to prove the i-equivalence for the next Piton step. The key is what might be called the "i-equivalence congruence" lemma, which states that i-equivalent states are produced by running the 'i' machine on i-equivalent states. But this relationship is not valid!

A computation on the 'i' machine can distinguish two i-equivalent states. More precisely, it is possible to step from two i-equivalent states to two states that are not i-equivalent. For example, suppose the instruction to be executed is **tpush_x**, which pushes onto the temporary stack the contents of the **x** register. Note that this instruction moves the contents of a hidden resource into a visible one. We call this *exposing* the hidden resource. The **tpush_x** instruction exposes **x** and if two i-equivalent states differ on **x**, then the execution of **tpush_x** produces non-i-equivalent states.

However, consider a basic block of i-code in which **x** is initialized from visible resources and then a **tpush_x** is done. The execution of that block of i-code in arbitrary i-equivalent states produces i-equivalent states, even though **tpush_x** in general exposes hidden resources. The basic block for each Piton instruction has the property that the hidden resources are all set before they are referenced—i.e., the hidden resources are all treated as temporaries within each basic block. The "i-equivalence congruence" lemma could be restated correctly by restricting our attention to i-code sequences with this property. However, it is difficult (though not impossible) to characterize such sequences.

The introduction of the 'r' machine allows us to leap over the problem. In one clock tick the 'r' machine carries out one Piton instruction and side-effects all the hidden resources appropriately. The r-equivalence congruence lemma is valid: it is impossible for an 'r' program to distinguish two r-equivalent states. In fact, the congruence lemma is straightforward to prove because the programming language supported by the 'r' machine provides no means of referencing the hidden resources. When we move down from 'r' to 'i'—running Piton instructions at the r-level and running the generated i-code at the i-level—we find the state diagram actually commutes—i.e., we do not need i-equivalence but can use equality—because the 'r' machine sets the hidden resources exactly as set by the generated i-code.

8.4.3. From I to M

We now move down one step further and relate the 'i' machine to the 'm' machine with a one-way correspondence theorem. Recall that the 'i' machine executes symbolic i-code on tagged data objects. The 'm' machine is FM9001 (except that it uses linear lists instead of ram trees). The memory and registers of 'm' contain only bit vectors. The relation between 'i' and 'm' is explained with 'i⇒m', which is the link-assembler.

THEOREM 3:

$$
\begin{aligned}
(\quad & (\textit{load-addr} \in \mathbf{N}) \\
\wedge \quad & (\text{i-psw}\,(\text{i}\,(i, n, \textit{load-addr})) = \text{'}\mathbf{run}) \\
\wedge \quad & (\text{i-word-size}\,(i) = \mathbf{32})) \\
\rightarrow (\quad & \text{i} \Rightarrow \text{m}\,(\text{i}\,(i, n, \textit{load-addr}), \textit{boot-lst}, \textit{load-addr}) \\
= \quad & \text{m}\,(\text{i} \Rightarrow \text{m}\,(i, \textit{boot-lst}, \textit{load-addr}), n))
\end{aligned}
$$

This theorem says that the binary image under 'i⇒m' of any non-erroneous i-level computation from some initial i-state can be alternatively obtained by running 'm' on the binary image of the initial i-state. Note that there is no sense of hidden resources or basic blocks of i-code here. We prove that the image of any i-code program executes correctly on 'm'.

Here, as in the 'p⇒r' proof, we see each machine taking the same number of steps. The proof of this one-way correspondence theorem thus breaks down immediately to the proof of the single step version for each machine, and that in turn breaks down to the case for each i-code instruction.

Below we exhibit the lemma for the i-code instruction **add_<tsp>{n}_x{n}**.

THEOREM:
- ((*load-addr* ∈ N)
- ∧   i-state-okp (*i, load-addr*)
- ∧   (i-psw (*i*) = **'run**)
- ∧   (i-word-size (*i*) = **32**)
- ∧   i-add_<tsp>{n}_x{n}-okp (*i*)
- ∧   (i-current-instruction (*i*) = **' (add_<tsp>{n}_x{n})**)))
- → (i⇒m (i-add_<tsp>{n}_x{n}-step (*i*), *boot-lst, load-addr*)
- =   m-step (i⇒m (*i, boot-lst, load-addr*)))

To prove this theorem, consider first the left-hand side of the conclusion. Symbolically expanding i-add_<tsp>{n}_x{n}-step (*i*) produces an i-state in which the topmost element of the temporary stack has been replaced by the tagged natural number obtained by summing the (untagged) old value on the stack and the (untagged) value of register **x**. Then symbolically evaluating the 'i⇒m' on that i-state produces a binary image. The key property of this image is that the bit vector at the address pointed to by register 3 (the **tsp** register) is the binary representation of the natural number sum described above.

Now consider the right-hand side of the conclusion, m-step (i⇒m (*i, boot-lst, load-addr*)). The first issue that must be faced in this proof is to determine what is the current instruction in the m-state i⇒m (*i, boot-lst, load-addr*), given that the current instruction in the i-state *i* is **(add_<tsp>{n}_x{n})**. The answer is that it is the bit vector obtained by calling 'link-instr-word' on **(add_<tsp>{n}_x{n})**, i.e., the bit vector corresponding to the assembly instruction **(add nil (tsp) x)**, which is the bit vector **000000111110000001001100000000100**. Given this current instruction, one can symbolically execute 'm-step', the single stepper for the 'm' machine. The result is a new m-state in which the bit vector at the address in register 3 (**0011**, the **tsp** register) is now the "binary-sum" of the old bit vector in that location and the contents of register 4 (**0100**, the **x** register). By "binary-sum" we mean the bit vector produced by the arithmetic-logical unit when the opcode is **0011**.

We see that the left-hand and right-hand m-states are equivalent if we simply know that the binary representation of the sum of two naturals is the binary-sum of the binary representations of the two naturals. This is true if the sum is representable, which is assured by the -okp predicate above.

In general the key lemmas that had to be proved to construct the 'i⇒m' level proofs had to do with

- the correspondence between the current instruction at the i-level and the current instruction at the m-level, as illustrated above;

- the commutativity of the fundamental data type operators and the bit-vector representations, as illustrated above; and

- the correspondence between name-offset address pairs into a structured memory segment at the i-level and the bit-vector addresses into a linear memory at the m-level.

The latter issue was actually involved in the illustration above. For example, when we fetched the current instruction of the m-state, we first inspected the program counter (register 0) and found there a bit-vector. That bit-vector, produced by 'i⇒m', was constructed by linking the data word which was the program counter at the i-level. That data word was an address pair that named a program and an offset. It was linked by computing a natural number indicating the analogous location in the linked memory and then converted to a bit-vector. When the 'm' machine fetches from that bit-vector address it gets the bit-vector produced by linking the instruction word found in the i-state at the address indicated by the original address pair.

It is here that we cross the bridge between the execute-only program space of the 'i' machine and the unrestricted (writeable) program space of 'm'. The statement "programs do not overwrite themselves" is not made explicitly. The one-way correspondence result is much stronger. Consider the region of the 'm' memory containing the link-assembled programs. Call that the "program segment" of the m-state. Consider the commutative diagram captured by the one-way correspondence theorem and the two paths to the final m-state. The 'm' program segment of the one is obtained by executing the 'i' machine and then link-assembling with 'p⇒r'. The 'm' program segment of the other is obtained by link-assembling and then executing the 'm' machine. But the two 'm' program segments must be identical because the two alternative m-states are identical. Thus, executing a link-assembled i-code instruction cannot change the 'm' program segment because the i-code instruction itself does not change the 'i' program segment.

Perhaps the most interesting instruction to prove correct at this level was the trivial i-code instruction (int-to-nat), which is the only instruction used in the basic block of the Piton instruction (int-to-nat). At the i-code (and Piton) level, (int-to-nat) pops an integer off the stack and pushes the corresponding natural, provided the integer was non-negative. At the machine code level, (int-to-nat) is mapped into a no-op.

The one-way correspondence lemma for that instruction is proved by the following steps. On the left-hand side, the i-level step retags the top of the stack from int to nat and then the 'i⇒m' maps it down, mapping the top of the stack as though it were an integer (since it is so tagged). On the right-hand side, 'i⇒m' maps down the original top of the stack as though it were a natural (since it is so tagged) and then the m-level step does nothing. The two m-states are equal because the bit-vector representing a small non-negative integer is the same as that representing the same natural.

8.4.4. From M to FM9001

The connection between 'm' and FM9001 is straightforward.

THEOREM 4:
proper-m-statep $(m) \rightarrow$ (m$\Rightarrow$fm9001 (m (m, n)) = fm9001 (m$\Rightarrow$fm9001 (m), n))

The function 'proper-m-statep' insures that the m-state has 16 registers in its register list and no more than 2^{32} items in its memory list. The proof is straightforward, given the isomorphism between positional addressing in linear lists and the tree-walk addressing in ram trees.

8.5. The Partial Inversion Lemmas

The stack of one-way correspondence lemmas takes us down from Piton to FM9001. It is the role of the partial inversion lemmas to bring the data segment back up. In our sketch of the proof on page 132 we used a diagram similar to the one in Figure 8-6 to indicate the role of the partial inversion lemmas. In fact, this diagram is misleading but serves the purpose of introducing the topic.

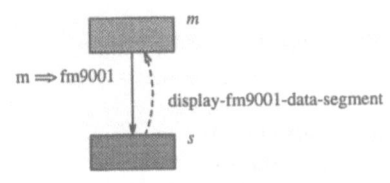

Figure 8-6: A Suggestive Partial Inversion

The idea in Figure 8-6 is as follows. Let m be an m-state and let s be its image under 'm$\Rightarrow$fm9001'. Then by applying 'display-fm9001-data-section' to s we can recover part of m. Which part? Intuitively, we might answer "we recover the data section of m." But m has no data section. Like s, its memory is an undifferentiated collection of bit vectors that we can interpret only through the link tables. So we consider first the structure of the link tables and the process of "unlinking," then we discuss how we display the memory in an m-state (i.e., how we partially invert 'i$\Rightarrow$m') and then we will relate that to the display of the FM9001 memory (i.e., the partial inversion of 'm$\Rightarrow$fm9001').

8.5.1. *Unlinking*

Unlinking is the process by which a bit vector is converted back into an abstract Piton object. To unlink a vector we must know the type of Piton object to be constructed and we must know the link tables used in the production of the original downloaded image. The type information is supplied by a type specification, which, recall, is supplied by the user who must know the final data types computed by the program. A type specification is a list that is isomorphic to the data segment to be constructed; the type specification names each area and lists the type of each element (thereby also indicating the length of each area).

There are four link tables. The program link table maps each program name to the absolute location of the beginning of the program. The label tables table maps each each program name to its label table, which is a table mapping the i-code labels of the program to the absolute position of the beginning of the basic block for that label. The third table, the user data link table, maps each global data area name to the absolute base address of the associated array. Finally, the system data link table maps each of the five system data area names to the absolute base address of the associated area. The four tables are illustrated in an extended example that starts on page 123.

Unlinking maps a desired type, a bit vector, and the link tables into a Piton object. The basic function is shown below.

DEFINITION:
unlink-data-word (*type*, *v*, *link-tables*)
=
case on *type*:
  case = **nat** **then** tag ('**nat**, v-to-nat (*v*))
  case = **int** **then** tag ('**int**, v-to-int (*v*))
  case = **bitv** **then** tag ('**bitv**, v-to-bitv (*v*))
  case = **bool** **then** tag ('**bool**, v-to-bool (*v*))
  case = **addr** **then** tag ('**addr**, v-to-addr (*v*, usr-data-links (*link-tables*)))
  case = **subr** **then** tag ('**subr**, v-to-subr (*v*, prog-links (*link-tables*)))
  case = **sys-addr**
  then tag ('**sys-addr**, v-to-sys-addr (*v*, sys-data-links (*link-tables*)))
  case = **pc** **then** tag ('**pc**, v-to-label (*v*, prog-label-tables (*link-tables*)))
  otherwise '(**unrecognized i-level type**) **endcase**

Observe that it case splits on the desired type and the uses the appropriate link tables to construct the object from the vector.

For example, if *type* is '**pc** and the vector *v* is the binary representation of the number 8029, and *link-tables* is as shown in the extended example on page 123, then 'unlink-data-word' uses the 'prog-label-tables', namely,

```
((main ((main prelude) . 8010)
       ((main . 0)      . 8012)
       ((main . 1)      . 8016)
       ((main . 2)      . 8018))
 (ptz ((ptz prelude)    . 8021)
      ((ptz . 0)        . 8023)
```

```
((ptz . 1)        . 8027)
((ptz . 2)        . 8029)
((ptz . 3)        . 8031))).
```

to construct the object ' (pc (ptz . 2)), by searching the table for the label that was assigned the address 8029.

The most basic theorem about unlinking is that it inverts linking. Formally, the theorem is

THEOREM: Unlinking Inverts Linking

$$((\textit{load-addr} \in N) \land \text{i-usr-data-objectp}(x, i) \land \text{i-state-okp}(i, \textit{load-addr}))$$
$$\rightarrow (\quad \text{unlink-data-word}(\text{type}(x),$$
$$\text{link-data-word}(x,$$
$$\text{i-link-tables}(i, \textit{load-addr}),$$
$$\text{i-word-size}(i)),$$
$$\text{i-link-tables}(i, \textit{load-addr}))$$
$$= \quad x)$$

Obviously, one must link and unlink with the same link tables. Less obviously, the theorem requires that the tables be those generated for the i-state in question. Consider, for example, the problem of linking ' (pc (ptz . 2)). What is the linker to do if the supplied link tables do not assign this pc an address?

8.5.2. Back to I from M

The function which displays the "m data segment," i.e., which recovers from an 'm' memory the data segment represented, is called 'display-m-data-segment'. It takes three arguments, an m-state, a type specification, and some link tables. It returns a Piton data segment. The type specification names the areas in the data segment to be reconstructed and specifies the type of each element in each area. The link tables are used for two purposes. First, each area named in the type specification is looked up in the user data link table to determine the absolute base address of the area. Then the memory of the m-state is scanned starting at the given base address and extending through as many words as were allocated to the area. Each word found in the memory is a 32-bit long bit vector and is unlinked according to the type specified for that element and the link tables. The unlinked objects are collected together into a list and the area name is added to the front. When such a list has been collected for each area, the result is returned.

The claim is that 'display-m-data-segment' partially inverts 'm⇒i'. The theorem is

THEOREM 5:

  (i-state-okp (i, *load-addr*)

  $\land$ (*load-addr* $\in$ **N**)

  $\land$ proper-i-usr-data-segmentp (i-usr-data-segment (i), i))

$\rightarrow$ (display-m-data-segment ($i \Rightarrow$ m (i, *boot-lst*, *load-addr*),

                               type-specification (i-usr-data-segment (i)),

                               i-link-tables (i, *load-addr*))

  = i-usr-data-segment (i))

and can be paraphrased as follows. Let i be a well-formed i-state for some numeric *load-addr*. Suppose the user data segment, *ds*, in i is proper, by which we mean that *ds* is a list of areas, each named by a unique literal atom and each containing only Piton data objects that are legal for the i-state i (e.g., every object tagged **addr** is of the form (**addr** (*name* . k)) where *name* is the name of an area in *ds* and k is a legal offset within that area). Let m be the m-state obtained by mapping i down with '$i \Rightarrow$m', *ts* be the type specification of *ds*, and let *tables* be the link tables created by 'i-link-tables'. Then display-m-data-segment (m, *ts*, *tables*) is in fact *ds*.

8.5.3. Back to M from FM9001

Having introduced 'display-m-data-segment' we now return to the FM9001 level. Consider an m-state m and its FM9001 image, s, under 'm$\Rightarrow$fm9001'. The FM9001 level display, applied to s, returns the same thing as the m-level display applied to m. Formally

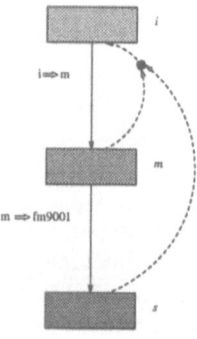

Figure 8-7: Partial Inversion of i$\Rightarrow$m and m$\Rightarrow$fm9001

THEOREM 6:
(proper-m-statep (m)

$\wedge$ plausible-data-link-tablep $(type\text{-}spec,$
                   length (m-mem (m)),
                   usr-data-links $(link\text{-}tables)))$

$\rightarrow$ (display-fm9001-data-segment (m$\Rightarrow$fm9001 (m), $type\text{-}spec$, $link\text{-}tables$)

=   display-m-data-segment $(m, type\text{-}spec, link\text{-}tables))$

The combination of the two inversion lemmas might be pictured as in Figure 8-7. Using the state labels of that diagram, displaying m produces the data segment of i and displaying s is the same as displaying m. Thus, displaying s produces the data segment of i.

However, note the second hypothesis. It requires that the user data link table be "plausible" for the type specification. This predicate requires that every area named in the type specification be assigned a base address in the user data link table and that the address be sufficiently small to allow the allocation of all the items in the area. If the link tables used in this theorem are "implausible" then the display functions will explore memory that does not exist.

8.6. The Correctness Proof

We wish to prove that Piton is correctly implemented on FM9001 by 'load'. The formula is shown below.

THEOREM: FM9001 Piton is Correct
(proper-p-statep (p_0)

$\wedge$ $(load\text{-}addr \in \mathbf{N})$

$\wedge$ p-loadablep $(p_0, load\text{-}addr)$

$\wedge$ (p-word-size (p_0) = $\mathbf{32}$)

$\wedge$ $(p_n = \mathrm{p}\,(p_0, n))$

$\wedge$ $(\neg$ errorp (p-psw $(p_n)))$

$\wedge$ $(ts = $ type-specification (p-data-segment $(p_n))))$

$\rightarrow$ (p-data-segment (p_n)

=   display-fm9001-data-segment (fm9001 (load $(p_0, boot\text{-}lst, load\text{-}addr)$,
                              fm9001-clock $(p_0, n))$,

                  ts,
                  link-tables $(p_0, load\text{-}addr)))$

In Figure 8-8 we diagram the proof of this formula. We will indicate briefly what the diagram in Figure 8-8 means in terms of formula manipulations. The manipulations treat the diagram more or less like a ladder: we start at the bottom and climb up to successive horizontal rungs.

Proof Sketch. We wish to prove that

p-data-segment (p (p_0, n))   [1]

is equal to

display-fm9001-data-segment (fm9001 (load $(p_0, ...)$, ...), ...).   [2]

Expanding the definition of 'load', [2] may be written

display-fm9001-data-segment ([3]
  fm9001 (m⇒fm9001 (i⇒m (r⇒i (p⇒r (p_0)), ...)), ...), ...).

By Theorem 4, we lift 'm⇒fm9001' out and convert the 'fm9001' to 'm'. The result
is

display-fm9001-data-segment ([4]
  m⇒fm9001 (m (i⇒m (r⇒i (p⇒r (p_0)), ...), ...)), ...).

By Theorem 6 we omit the 'm⇒fm9001' step, converting 'display-fm9001-data-
segment' to 'display-m-data-segment'.

display-m-data-segment (m (i⇒m (r⇒i (p⇒r (p_0)), ...), ...), ...).   [5]

By Theorem 3 we lift out 'i⇒m' and convert 'm' to 'i', obtaining

display-m-data-segment (i⇒m (i (r⇒i (p⇒r (p_0)), ...), ...), ...),   [6]

which we further simplify to

i-usr-data-segment (i (r⇒i (p⇒r (p_0)), ...)),   [7]

by the partial inversion result for 'i⇒m', Theorem 5. Lifting 'r⇒i' out and converting 'i' to 'r' with Theorem 2 produces

i-usr-data-segment (r⇒i (r (p⇒r (p_0), n))).   [8]

But 'r⇒i' does not change the data segment. More precisely, the 'i-usr-data-
segment' of the image of any r under 'r⇒i' is just the 'r-usr-data-segment' of that r.
This is obvious from the definition of 'r⇒i' and is so labeled in the diagram of
Figure 8-8. Thus, [8] is

r-usr-data-segment (r (p⇒r (p_0), n)).   [9]

By Corollary 1, [9] is

r-usr-data-segment (p⇒r (p (p_0, n))).   [10]

But 'p⇒r' does not change the data segment of its input. Hence, it is obvious that
[10] is

p-data-segment (p (p_0, n)),   [11]

which is [1], as desired. **Q.E.D.**

Why is this a "proof sketch" rather than a "proof?" One reason is that we have
ignored the hypotheses of the Theorems we used. Another reason is that the ellipses
above hide some fundamental problems.

We consider three of the omitted hypotheses to illustrate the level of detail
brushed aside in our proof sketch.

In the production of line [4] from line [3] we used

THEOREM 4:
proper-m-statep (m) → (m⇒fm9001 (m (m, n)) = fm9001 (m⇒fm9001 (m), n))

In the application in question, m, above, is instantiated with i⇒m (r⇒i (p⇒r (p_0)),
...) so the right hand side of Theorem 4 matches line [3]. That is, m is the m_0 of
Figure 8-8. How do we know that m_0 is a 'proper-m-statep'? The answer is that we

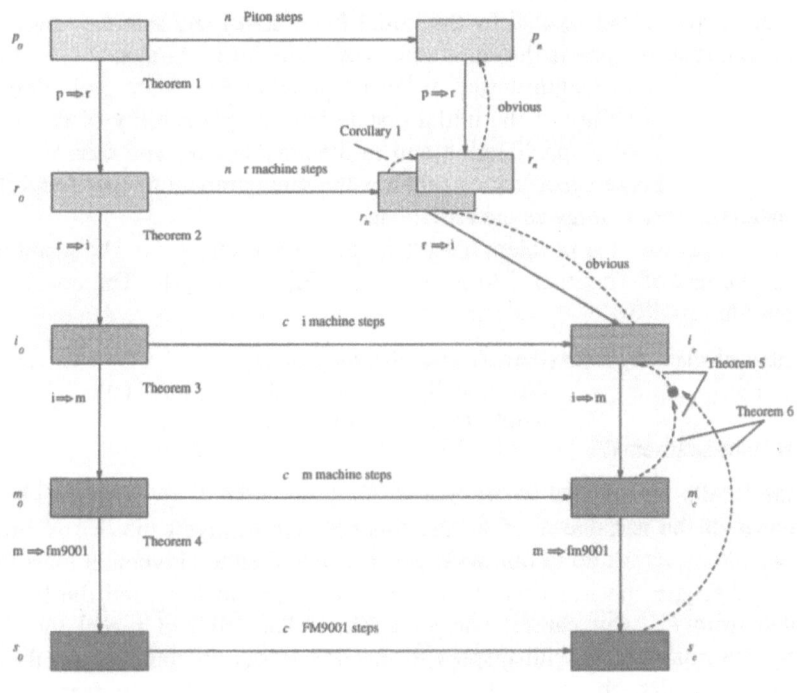

Figure 8-8: Diagram of the Piton Correctness Proof

know p_0 is a loadable proper p-state and we can prove that the "proper state" notions at the various levels are preserved under the downward mappings. We have previously mentioned only one of these preservation lemmas, but they must each be proved to complete the proof sketch above.

Similarly, when we applied Theorem 3 to line [5] we omitted mention of the hypothesis that the 'i-psw' of the final i-state (i_c of Figure 8-8) is **'run**. In the instance of Theorem 3 in question the final i-state is i $(r{\Rightarrow}i\,(p{\Rightarrow}r\,(p_0))$, c, *load-addr*), for a certain 'clock' expression c. We know the 'i-psw' of this state is **'run** because we know that the 'p-psw' of p_n is not **'error** and the one-way correspondence theorems for 'p⇒r' and 'r⇒i' (Theorems 1 and 2) can be used to show that the 'i-psw' of i_c must thus be **'run**.

Finally, consider the requirement of Theorem 6 that the user data link table be "plausible" for the type specification (page 157). When Theorem 6 is used at line [4] to produce line [5], the instance of m is the final m-level state, m_c of Figure 8-8. We must show that the data link table is plausible for the length of the memory in that final state. Essentially all we have to work with is that p_0 is a loadable proper p-state. Note immediately that the "source" assumption is about a top-level initial state and the "destination" goal is about a low-level final state. Thus, we must pass the information both across and down. The key lemmas establish that Piton never changes the length of a data area (thus the amount of space needed for a final Piton state is the same as that needed by the initial Piton state) and that the length of the memory in a final m-state is the same as it was in the initial m-state— i.e., that 'm' memory is of constant length during m-level computations. With such lemmas we can convert the loadability of the initial p-state into the plausibility of the data link table for the initial type specification and m-memory length, and then we can observe that the final type specification requires the same amount of space and that the final memory is just as large as the initial one.

Finally, we illustrate a problem hidden by our use of ellipses in the proof sketch. Consider the use of Theorem 5 to go from line [6] to line [7]. The conclusion of Theorem 5 is

display-m-data-segment $(i \Rightarrow m$ $(i,$ *boot-lst, load-addr*$)$,

$\qquad$ type-specification (i-usr-data-segment (i)),

$\qquad$ i-link-tables $(i,$ *load-addr*$))$

$=$ $\quad$ i-usr-data-segment (i)

Theorem 5 tells us how to invert 'i$\Rightarrow$m' on some state i. Note particularly the occurrences of the variable i. If we use this theorem to invert the 'i$\Rightarrow$m' image of the final state, i_c, as we do in our main proof above, then the inversion must be with respect to the type specification of the user data segment in i_c and the link tables computed from i_c. But careful consideration of line [6] will reveal that we are inverting the image of i_c with respect to the link tables computed from the initial state! In particular, the final ellipsis in line [6] comes directly from the main theorem and stands for link-tables $(p_0,$ *load-addr*$)$, which, by definition, is i-link-tables $(r \Rightarrow i$ $(p \Rightarrow r$ $(p_0))$, *load-addr*$)$ or i-link-tables $(i_0,$ *load-addr*$)$.

Thus, to use Theorem 5 as we have, we must show that the link tables computed for i_0 are the same as those computed for i_n. The heart of that proof is that the i-machine does not include any instruction that changes the length of any data area.

It is the overwhelming number of such details that make proofs about software mind-numbing. That in turn is one of the reasons the "social process" of mathematics (page 6) is an ineffective means of checking such proofs. Fortunately, the Piton proof has been mechanically checked.

Summary of Piton Instructions

In this appendix we describe each of the current Piton instructions informally in the manner of Chapter 3. This appendix is intended for use by the Piton programmer.

The instructions are listed in alphabetical order. Unless otherwise indicated, every instruction increments the program counter by one so that the next instruction to be executed is the instruction following the current one in the current subroutine. All references to "the stack" refer to the temporary stack unless otherwise specified. When we say "push" or "pop" without mentioning a particular stack we mean to push or pop the temporary stack.

Each instruction description has four parts. First we exhibit the syntactic form of the instruction, e.g., (**jump** *lab*). Next, under the heading "*Syn*" (for "Syntactic Restrictions") we give a page number and then informally characterize the syntactic constraints on the instruction. For example, for (**jump** *lab*) we say "*lab* is a label in the containing program." The page number is the page on which the formalization of this syntactic constraint is found, e.g., the page on which 'proper-p-jump-instructionp' is defined. In general, the syntactic well-formedness criteria for the Piton instruction named *name* are checked by the function 'proper-p-*name*-instructionp'. The next part of the informal description is headed by "*Pre*" (for "Precondition"). There we list a page number and give an informal description of the runtime conditions that must be satisfied in order to execute the instruction without error. For example, for the **pop** instruction we say "there is at least one item on the stack." The page number indicates where the restriction is expressed formally. In general, the runtime conditions on an instruction named *name* are checked by the function 'p-*name*-okp'. Finally, under the heading "*Effects*" we give a page number and an informal description of how the instruction changes the Piton state. The page number indicates where the "step" function for that instruction is defined. The step function for the instruction named *name* is 'p-*name*-step'.

(**add-addr**)   *Syn* (page 227): None. *Pre* (page 185): There is a natural, *n*, on top of the stack and a data address, *a*, immediately below it. The result of incrementing *a* by *n* is a legal data address. *Effects* (page 185): Pop twice and then push the data address obtained by incrementing *a* by *n*.

(add-int)   *Syn* (page 228): None. *Pre* (page 186): There is an integer, i, on top of the stack and an integer, j, immediately below it. The integer sum of i and j is representable. *Effects* (page 186): Pop twice and then push the integer sum of i and j.

(add-int-with-carry)
         Syn (page 228): None. *Pre* (page 186): There is an integer, i, on top of the stack, an integer, j, immediately below it, and a Boolean, c, below that. *Effects* (page 187): Pop three times. Let k be **1** if c is **t** and **0** otherwise. Let *sum* be the integer sum $i + j + k$. If *sum* is representable in the word size, w, of this p-state, push the Boolean **f** and then the integer *sum*; if *sum* is not representable and is negative, push the Boolean **t** and the integer *sum* $+ 2^w$; if sum is not representable and positive, push the Boolean **t** and the integer *sum* $- 2^w$.

(add-nat)   *Syn* (page 228): None. *Pre* (page 187):There is a natural, i, on top of the stack and a natural, j, immediately below it. The natural $j + i$ is representable. *Effects* (page 187): Pop twice and then push the natural $j + i$.

(add-nat-with-carry)
         Syn (page 228): None. *Pre* (page 188): There is a natural, i, on top of the stack, a natural, j, immediately below it, and a Boolean, c, immediately below that. *Effects* (page 188): Pop three times. Let k be **1** if c is **t** and **0** otherwise. Let sum be the natural $i + j + k$. If *sum* is representable in the word size, w, of this p-state, push the Boolean **f** and then natural *sum*; if *sum* is not representable, push the Boolean **t** and the natural *sum* $- 2^w$.

(add1-int)   *Syn* (page 228): None. *Pre* (page 188): There is an integer, i, on top of the stack and the integer $i + 1$ is representable. *Effects* (page 189): Pop once and then push the integer $i + 1$.

(add1-nat)   *Syn* (page 228): None. *Pre* (page 189): There is a natural, i, on top of the stack and the natural $i + 1$ is representable. *Effects* (page 189): Pop once and then push the natural $i + 1$.

(and-bitv)   *Syn* (page 228): None. *Pre* (page 189): There is a bit vector, v_1, on top of the stack and a bit vector, v_2, immediately below it. *Effects* (page 190): Pop twice and then push the bit vector result of the componentwise conjunction of v_1 and v_2.

(and-bool)   *Syn* (page 228): None. *Pre* (page 190): There is a Boolean, b_1, on top of the stack and a Boolean, b_2, immediately below it. *Effects* (page 190): Pop twice and then push the Boolean conjunction of b_1 and b_2.

(call *subr***)**   *Syn* (page 228): *Subr* is the name of a program in the program segment. *Pre* (page 190): Suppose that *subr* has n formal variables and k temporary variables. Then the temporary stack must contain at least n items and the control stack must have at least 2 $+ n + k$ free slots. *Effects* (page 191): Transfer control to the first instruction in the body of *subr* after removing the topmost n elements from the temporary stack and constructing a new frame on the control stack. In the new frame the formals of *subr* are bound to the n elements removed from the temporary stack, in reverse order, the temporaries of *subr* are bound to their declared initial values, and the return program counter points to the instruction after the **call**.

(deposit)   *Syn* (page 229): None. *Pre* (page 192): There is a data address, a, on top of the stack and an arbitrary object, *val*, immediately below it. *Effects* (page 192): Pop twice and then deposit *val* into the location addressed by a.

(deposit-temp-stk)
        Syn (page 229): None. *Pre* (page 192): There is a natural number, n, on top of the stack and some object, *val*, immediately below it. Furthermore, n is less than the length of the stack after popping two elements. *Effects* (page 192): Pop twice and then deposit *val* at the nth position in the temporary stack, where positions are enumerated from 0 starting at the bottom.

(div2-nat)   *Syn* (page 229): None. *Pre* (page 192): There is a natural, i, on top of the stack and room to push at least one more item. *Effects* (page 193): Pop once and then push the natural floor of the quotient of i divided by 2 and then push the natural $i \bmod 2$.

(eq)   *Syn* (page 229): None. *Pre* (page 193): The temporary stack contains at least two items and the top two are of the same type. *Effects* (page 193): Pop twice and then push the Boolean **t** if they are the same and the Boolean **f** if they are not.

(fetch)   *Syn* (page 229): None. *Pre* (page 193): There is a data address, a, on top of the stack. *Effects* (page 193): Pop once and then push the contents of address a.

(fetch-temp-stk)

> *Syn* (page 229): None. *Pre* (page 194): There is a natural number, n, on top of the stack and n is less than the length of the stack. *Effects* (page 194): Let *val* be the nth element of the stack, where elements are enumerated from 0 starting at the bottom-most element. Pop once and then push *val*.

(int-to-nat) *Syn* (page 232): None. *Pre* (page 200): There is a non-negative integer, i, on top of the stack. *Effects* (page 200): Pop and then push the natural i.

(jump *lab***)**   *Syn* (page 233): *Lab* is a label in the containing program. *Pre* (page 202): None. *Effects* (page 202): Jump to *lab*.

(jump-case lab_0 lab_1 ... lab_n**)**

> *Syn* (page 233): Each of the lab_i is a label in the containing program. *Pre* (page 201): There is a natural, i, on top of the stack and $i \leq n$. *Effects* (page 201): Pop once and then jump to lab_i.

(jump-if-temp-stk-empty *lab***)**

> *Syn* (page 233): *Lab* is a label in the containing program. *Pre* (page 201): None. *Effects* (page 201): Jump to *lab* if the temporary stack is empty.

(jump-if-temp-stk-full **lab)**

> *Syn* (page 233): *Lab* is a label in the containing program. *Pre* (page 202): None. *Effects* (page 202): Jump to *lab* if the temporary stack is full.

(locn *lvar***)**   *Syn* (page 233): *Lvar* is a local variable of the containing program. *Pre* (page 203): The value, n, of *lvar* is a natural number less than the number of locals of the current program. *Effects* (page 203): Push the value of the nth local variable of the current program.

(lsh-bitv)   *Syn* (page 233): None. *Pre* (page 203): There is a bit vector, v, on top of the stack. *Effects* (page 203): Pop once and then push the bit vector result of left shifting each bit of v, bringing in a 0 on the right.

(lt-addr)   *Syn* (page 233): None. *Pre* (page 203): There is a data address,

a_1, on top of the stack and a data address, a_2, immediately below it. a_1 and a_2 address the same data area. *Effects* (page 204): Pop twice and then push the Boolean **t** if $a_2 < a_1$ (i.e., the position addressed by a_2 is to the left of that addressed by a_1) and push the Boolean **f** otherwise.

(lt-int)   *Syn* (page 233): None. *Pre* (page 204): There is an integer, i, on top of the stack and an integer, j, immediately below it. *Effects* (page 204): Pop twice and then push the Boolean **t** if $j < i$ and the Boolean **f** otherwise.

(lt-nat)   *Syn* (page 233): None. *Pre* (page 204): There is a natural, i, on top of the stack and a natural, j, immediately below it. *Effects* (page 205): Pop twice and then push the Boolean **t** if $j < i$ and the Boolean **f** otherwise.

(mult2-nat)   *Syn* (page 233): None. *Pre* (page 205): There is a natural, i, on top of the stack and $2 \times i$ is representable. *Effects* (page 205): Pop once and then push the natural $2 \times i$.

(mult2-nat-with-carry-out)
                   Syn (page 234): None. *Pre* (page 205): There is a natural, i, on top of the stack and room to push at least one more item. *Effects* (page 206): Pop once. Then, if the natural $2 \times i$ is representable, push the Boolean **f** and then the natural $2 \times i$. Otherwise, push the Boolean **t** and the natural $(2 \times i) - 2^w$, where w is the word size of this p-state.

(neg-int)   *Syn* (page 234): None. *Pre* (page 206): There is an integer, i, on top of the stack and $-i$ is representable. *Effects* (page 206): Pop once and then push the integer $-i$.

(no-op)   *Syn* (page 234): None. *Pre* (page 206): None. *Effects* (page 207): Do nothing; continue execution.

(not-bitv)   *Syn* (page 234): None. *Pre* (page 207): There is a bit vector, v, on top of the stack. *Effects* (page 207): Pop once and then push the bit vector result of the componentwise logical negation of v.

(not-bool)   *Syn* (page 234): None. *Pre* (page 207): There is a Boolean, b, on top of the stack. *Effects* (page 207): Pop once and then push the Boolean negation of b.

(or-bitv)   *Syn* (page 234): None. *Pre* (page 208): There is a bit vector, v_1, on top of the stack and a bit vector, v_2, immediately below it. *Effects* (page 208): Pop twice and then push the bit vector result of the componentwise disjunction of v_1 and v_2.

(or-bool)   *Syn* (page 234): None. *Pre* (page 209): There is a Boolean, b_1, on top of the stack and a Boolean, b_2, immediately below it. *Effects* (page 209): Pop twice and then push the Boolean disjunction of b_1 and b_2.

(pop)   *Syn* (page 234): None. *Pre* (page 212): There is at least one item on the stack. *Effects* (page 212): Pop and discard the top of the stack.

(pop* *n*)   *Syn* (page 234): *N* is a natural number. Note: **(pop* 3)** is well formed; **(pop* (nat 3))** is not. *Pre* (page 209): there are at least *n* items on the stack. *Effects* (page 209): Pop and discard the topmost *n* items.

(pop-call)   *Syn* (page 234): None. *Pre* (page 210): A subroutine name, *subr*, is on top of the stack and, after removing that name, it is legal to **call** *subr* (i.e., sufficient arguments are on the temporary stack and the control stack has room for the new frame). *Effects* (page 210): Pop once and then execute **(call** *subr***)**.

(pop-global *gvar***)**

          Syn (page 234): *Gvar* is a global variable, i.e., the name of a data area in the data segment of the containing p-state. *Pre* (page 210): There is an object, *val*, on top of the stack. *Effects* (page 210): Pop and assign *val* to the (0th position of the array associated with the) global variable *gvar*.

(pop-local *lvar***)**

          Syn (page 234): *Lvar* is a local variable of the containing program. *Pre* (page 211): There is an object, *val*, on top of the stack. *Effects* (page 211): Pop and assign *val* to the local variable *lvar*.

(pop-locn *lvar***)**

          Syn (page 235): *Lvar* is a local variable of the containing program. *Pre* (page 211): The value, *n*, of *lvar* is less than the number of local variables of the current program and there is an object, *val*, on top of the stack. *Effects* (page 211): Pop and assign *val* to the nth local variable.

(popj)   *Syn* (page 235): None. *Pre* (page 212): There is a program
                  counter object, *pc*, on top of the stack and *pc* addresses the
                  current program. *Effects* (page 212): Pop once and then transfer
                  control to *pc*.

(popn)   *Syn* (page 235): None. *Pre* (page 213): There is a natural, *n*, on
                  top of the stack and there are at least *n* items on the stack below
                  it. *Effects* (page 213): Pop and discard *n* + **1** items. Thus, in
                  order to pop *n* items off the stack, you should push *n* onto the
                  stack and execute **(popn)**.

(push-constant *const***)**
                  Syn (page 235): *Const* is either a legal Piton object in the con-
                  taining p-state, the atom **pc**, or a label in the containing program.
                  Pre (page 213): There is room to push at least one item. *Effects*
                  (page 213): If *const* is a Piton object, push *const*; if *const* is the
                  atom **pc**, push the program counter of the next instruction; other-
                  wise push the program counter corresponding to the label *const*.

(push-ctrl-stk-free-size)
                  Syn (page 236): None. *Pre* (page 214): There is room to push
                  at least one item. *Effects* (page 214): Push the natural number
                  indicating how many more cells can be created on the control
                  stack before the maximum control stack size is exceeded.

(push-global *gvar***)**
                  Syn (page 236): *Gvar* is a global variable, i.e., the name of a
                  data area in the data segment of the containing p-state. *Pre* (page
                  214): There is room to push at least one item. *Effects* (page
                  214): Push the value of the global variable *gvar*, i.e., the con-
                  tents of position **0** in the array associated with *gvar*.

(push-local *lvar***)**
                  Syn (page 236): *Lvar* is a local variable of the containing
                  program. *Pre* (page 215): There is room to push at least one
                  item. *Effects* (page 215): Push the value of the local variable
                  lvar.

(push-temp-stk-free-size)
                  Syn (page 236): None. *Pre* (page 215): There is room to push
                  at least one item. *Effects* (page 215): Push the natural number
                  indicating how many more cells can be created on the temporary
                  stack before the maximum temporary stack size is exceeded.

(push-temp-stk-index *n*)

Syn (page 236): *N* is a natural number. Note: *n* here must not be tagged; **(push-temp-stk-index 3)** is well formed and **(push-temp-stk-index (nat 3))** is not. *Pre* (page 216): *N* is less than the length of the temporary stack and there is room to push at least one item. *Effects* (page 216): Push the natural number (*length* − *n*) − **1** where *length* is the current length of the temporary stack. Note: We permit the temporary stack to be accessed randomly as an array. The elements in the stack are enumerated from **0** starting at the *bottom-most* so that pushes and pops do not change the positions of undisturbed elements. This instruction converts from a topmost-first enumeration to our enumeration. That is, it pushes onto the temporary stack the index of the element *n* removed from the top. See also **fetch-temp-stk** and **deposit-temp-stk**.

(pushj *lab*)

Syn (page 236): *Lab* is a label in the containing program. *Pre* (page 216): There is room to push at least one item. *Effects* (page 216): Push the program counter addressing the next instruction and then jump to *lab*.

(ret)

Syn (page 236): None. *Pre* (page 217): None. *Effects* (page 217): If the control stack contains only one frame (i.e., if the current invocation is the top-level entry into Piton) **halt** the machine. Otherwise, set the program counter to the return program counter in the topmost frame of the control stack and pop that frame off the control stack.

(rsh-bitv)

Syn (page 236): None. *Pre* (page 217): There is a bit vector, *v*, on top of the stack. *Effects* (page 217): Pop once and then push the bit vector result of right shifting each bit in *v*, bringing in a **0** on the left.

(set-global *gvar*)

Syn (page 237): *Gvar* is a global variable, i.e., the name of a data area in the data segment of the containing p-state. *Pre* (page 218): There is an object, *val*, on top of the stack. *Effects* (page 218): Assign *val* to the (0th position of the array associated with the) global variable *gvar*. The stack is not popped.

(set-local *lvar*)

Syn (page 237): *Lvar* is a local variable in the containing program. *Pre* (page 218): There is an object, *val*, on top of the stack. *Effects* (page 218): Assign *val* to the local variable *lvar*. The stack is not popped.

(sub-addr)   *Syn* (page 237): None. *Pre* (page 219): There is a natural, n, on top of the stack and a data address, a, immediately below it. The result of decrementing a by n is a legal data address. *Effects* (page 219): Pop twice and then push the data address obtained by decrementing a by n.

(sub-int)   *Syn* (page 237): None. *Pre* (page 219): There is an integer, i, on top of the stack and an integer, j, immediately below it. The integer $j - i$ is representable. *Effects* (page 220): Pop twice and then push the integer $j - i$.

(sub-int-with-carry)
   Syn (page 237): None. *Pre* (page 220): There is an integer, i, on top of the stack, an integer, j, immediately below it, and a Boolean, c, below that. *Effects* (page 220): Pop three times. Let k be **1** if c is **t** and **0** otherwise. Let *diff* be the integer $j - (i + k)$. If *diff* is representable in the word size, w, of this p-state, push the Boolean **f** and the integer *diff*; if *diff* is not representable and is negative, push the Boolean **t** and the integer $diff + 2^w$; if *diff* is not representable and is positive, push the Boolean **t** and the integer $diff - 2^w$.

(sub-nat)   *Syn* (page 237): None. *Pre* (page 221): There is a natural, i, on top of the stack and natural, j, immediately below it. Furthermore, $j \geq i$. *Effects* (page 221): Pop twice and then push the natural $j - i$.

(sub-nat-with-carry)
   Syn (page 237): None. *Pre* (page 222): There is a natural, i, on top of the stack, a natural, j, immediately below it, and a Boolean, c, immediately below that. *Effects* (page 222): Pop three times. Let k be **1** if c is **t** and 0 otherwise. If $j \geq i + k$, then push the Boolean **f** and the natural $j - (i + k)$. Otherwise, push the Boolean **t** and the natural $2^w - ((i + k) - j)$, where w is the word size of this p-state.

(sub1-int)   *Syn* (page 238): None. *Pre* (page 222): There is an integer, i, on top of the stack and the integer $i - 1$ is representable. *Effects* (page 222): Pop once and then push the integer $i - 1$

(sub1-nat)   *Syn* (page 238): None. *Pre* (page 223): There is a non-zero natural, i, on top of the stack. *Effects* (page 223): Pop and then push the natural $i - 1$.

(test-bitv-and-jump *test lab***)**

Syn (page 238): *Test* is either **all-zero** or **not-all-zero** and *lab* is a label in the containing program. *Pre* (page 224): There is a bit vector, v, on top of the stack. *Effects* (page 224): Pop once and then jump to *lab* if *test* is satisfied, as indicated below.

test	condition tested
all-zero	every component of v is **0**
not-all-zero	some component of v is **1**

(test-bool-and-jump *test lab***)**

Syn (page 238): *Test* is either **t** or **f** and *lab* is a label in the containing program. *Pre* (page 224): There is a Boolean, b, on top of the stack. *Effects* (page 225): Pop once and then jump to *lab* if *test* is satisfied, as indicated below.

test	condition tested
t	$b = \text{t}$
f	$b = \text{f}$

(test-int-and-jump *test lab***)**

Syn (page 238): *Test* is one of **neg, not-neg, zero, not-zero, pos** or **not-pos**, and *lab* is a label in the containing program. *Pre* (page 225): There is an integer, i, on top of the stack. *Effects* (page 225): Pop once and then jump to *lab* if *test* is satisfied, as indicated below.

test	condition tested
neg	$i < 0$
not-neg	$i \geq 0$
zero	$i = 0$
not-zero	$i \neq 0$
pos	$i > 0$
not-pos	$i \leq 0$

(test-nat-and-jump *test lab***)**

Syn (page 238): *Test* is either **zero** or **not-zero** and *lab* is a label in the containing program.[15] *Pre* (page 225): There is a natural, n, on top of the stack. *Effects* (page 226): Pop once and then jump to *lab* if *test* is satisfied, as indicated below.

---

[15]Technically, *test* may be anything whatsoever. If it is not **zero** it is treated as though it were **not-zero**.

test	condition tested
zero	$n = 0$
not-zero	$n \neq 0$

(xor-bitv)   *Syn* (page 239): None. *Pre* (page 226): There is a bit vector, v_1, on top of the stack and a bit vector, v_2, immediately below it. *Effects* (page 226): Pop twice and then push the bit vector result of the componentwise exclusive-or of v_1 and v_2.

Appendix II

The Formal Definition of Piton

This appendix contains all of the formulas involved in the formal definition of Piton. The appendix is divided into two sections. The second section is simply a listing, in alphabetical order, of all of the definitions involved. These definitions are indexed and exhaustively cross-indexed in the Index of this book. The first section here is a guide to the second section. It briefly mentions each of the important "entry points" into the list of definitions.

II.1. A Guide to the Formal Definition of Piton

II.1.1. Proper P-States

P-states are formally represented by the 'p-state' shell

SHELL DEFINITION:
Add the shell 'p-state' of 9 arguments, with
recognizer function symbol 'p-statep', and
accessors 'p-pc',
        'p-ctrl-stk', 'p-temp-stk',
        'p-prog-segment', 'p-data-segment',
        'p-max-ctrl-stk-size', 'p-max-temp-stk-size',
        'p-word-size', and 'p-psw'.

The proper p-states are described by 'proper-p-statep',

DEFINITION:

proper-p-statep (p)

=

p-statep (p)	; (1)
$\wedge$ p-objectp-type $('\text{\textbf{pc}}, \text{p-pc}(p), p)$	; (2)
$\wedge$ listp (p-ctrl-stk (p))	; (3)
$\wedge$ proper-p-framep (top (p-ctrl-stk (p)), area-name (p-pc (p)), p)	; (4)
$\wedge$ proper-p-ctrl-stkp (pop (p-ctrl-stk (p)),	; (5)

$$\text{area-name (ret-pc (top (p-ctrl-stk}(p)))),$$
$$p)$$

$\wedge$ (p-max-ctrl-stk-size (p) $\geq$ p-ctrl-stk-size (p-ctrl-stk (p)))	; (6)
$\wedge$ proper-p-temp-stkp (p-temp-stk (p), p)	; (7)
$\wedge$ (p-max-temp-stk-size (p) $\geq$ length (p-temp-stk (p)))	; (8)
$\wedge$ proper-p-prog-segmentp (p-prog-segment (p), p)	; (9)
$\wedge$ proper-p-data-segmentp (p-data-segment (p), p)	; (10)
$\wedge$ (p-max-ctrl-stk-size (p) $\in \mathbf{N}$)	; (11)
$\wedge$ (p-max-temp-stk-size (p) $\in \mathbf{N}$)	; (12)
$\wedge$ (p-word-size (p) $\in \mathbf{N}$)	; (13)
$\wedge$ (p-max-ctrl-stk-size (p) $< 2^{\text{p-word-size}(p)}$)	; (14)
$\wedge$ (p-max-temp-stk-size (p) $< 2^{\text{p-word-size}(p)}$)	; (15)
$\wedge$ (0 $<$ p-word-size (p))	; (16)

The sixteen conjuncts of this definition may be paraphrased as follows. (1) p is a p-state (see 'p-state', page 218); (2) the program counter of p is a legal Piton object of type **PC** (see 'p-objectp-type', page 208); (3) the control stack is non-empty; (4) the topmost frame of the control stack is a proper frame for the current program counter and state (see 'proper-p-framep', page 229); (5) the rest of the control stack is similarly proper (see 'proper-p-ctrl-stkp', page 229); (6) the "size" of the control stack does not exceed the specified limit (see 'p-ctrl-stk-size', page 191); (7) the temporary stack is proper in the current state, which means it consists entirely of legal Piton objects (see 'proper-p-temp-stkp', page 238); (8) the length of the temporary stack does not exceed the specified limit; (9) the program segment is well-formed (see 'proper-p-prog-segmentp', page 235 and below); (10) the data segment is well-formed (see 'proper-p-data-segmentp', page 229); (11-13) the maximum control stack size, the maximum temporary stack size, and the word size, w, are all natural numbers; (14) the maximum control stack size is less than 2^w; (15) the maximum temporary stack size is less than 2^w; and (16) w is greater than 0.

The reader is encouraged to pursue each of the references above. We will briefly elaborate on the definition of 'proper-p-prog-segmentp'.

DEFINITION:

proper-p-prog-segmentp $(segment, p)$

=

if nlistp $(segment)$ **then** $segment = $ **nil**
  else   proper-p-programp (car $(segment)$, p)
        $\wedge$ proper-p-prog-segmentp (cdr $(segment)$, p) **endif**

This function is recursive. It requires that the (program) $segment$ on which it was called be a list ending in **nil**, every element of which is a 'proper-p-programp' with respect to the current p-state.

Looking up 'proper-p-programp' in the index, we find that it is defined on page 235, as follows:

DEFINITION:
proper-p-programp *(prog, p)*
=
    litatom (name *(prog)*)
∧ all-litatoms (formal-vars *(prog)*)
∧ proper-p-temp-var-dclsp (temp-var-dcls *(prog)*, *p*)
∧ proper-p-program-bodyp (program-body *(prog)*, name *(prog)*, *p*)

This function checks that the name of the program is a literal atom, the formal variable field contains a list of literal atoms, the temporary variable declaration field contains a proper list of declarations (in particular, each specifies a literal atom as a variable and a legal Piton object as its initial value), and the body of the program is proper.

On page 235 we see

DEFINITION:
proper-p-program-bodyp *(lst, name, p)*
=
    listp *(lst)*
∧ proper-labeled-p-instructionsp *(lst, name, p)*
∧ fall-off-proofp *(lst)*

That is, a proper Piton program body is a non-empty list of properly labeled Piton instructions and has the "fall off proof" property. The latter term means that the last instruction in the list is an unconditional transfer of control—i.e., it is not possible to "fall off the end" of the list by executing the instructions.

We continue our dive by looking at 'proper-labeled-p-instructionsp'

DEFINITION:
proper-labeled-p-instructionsp *(lst, name, p)*
=
if nlistp *(lst)* **then** *lst* = **nil**
 else   legal-labelp (car *(lst)*)
       ∧ proper-p-instructionp (unlabel (car *(lst)*), *name, p*)
       ∧ proper-labeled-p-instructionsp (cdr *(lst)*, *name, p*) **endif**

This recursive function checks that each element of the body is legally labeled (if at all) and that the result of unlabeling the element is a 'proper-p-instructionp' as defined on page 230

DEFINITION:
proper-p-instructionp (ins, $name$, p)
=

    properp (ins)
$\land$ **case on** car (ins):
    case = call
    then proper-p-call-instructionp (ins, $name$, p)
    case = ret
    then proper-p-ret-instructionp (ins, $name$, p)
    case = locn
    then proper-p-locn-instructionp (ins, $name$, p)
    case = push-constant
    then proper-p-push-constant-instructionp (ins, $name$, p)

    ...
    case = or-bool
    then proper-p-or-bool-instructionp (ins, $name$, p)
    case = and-bool
    then proper-p-and-bool-instructionp (ins, $name$, p)
    case = not-bool
    then proper-p-not-bool-instructionp (ins, $name$, p)
    otherwise f endcase

Observe that 'proper-p-instructionp' splits on the opcode of the Piton instruction and for each opcode calls the appropriate predicate to check that the instruction is well-formed.

Consider, for example, the **push-constant** instruction. The predicate that checks whether it is well-formed is named 'proper-p-push-constant-instructionp'

DEFINITION:
proper-p-push-constant-instructionp (ins, $name$, p)
=

    (length (ins) = **2**)
$\land$ (p-objectp (cadr (ins), p)
    $\lor$ (cadr (ins) = **'pc**)
    $\lor$ find-labelp (cadr (ins),
                             program-body (definition ($name$, p-prog-segment (p)))))

This predicate checks that the instruction is of length 2, i.e., has the form **(push-constant** x**)**, and that x is either a legal Piton object in the state containing the instruction, or is the atom **'pc**, or is a label in the program containing the instruction.

In general, to determine the syntactic restrictions on an instruction named $opcode$, look at the definition of the function named 'proper-p-$opcode$-instructionp'.

This completes our brief tour through the definition of proper p-states. The serious student of Piton should use the index to trace out the entire tree of definitions.

II.1.2. The Piton Interpreter

The Piton interpreter is the function named 'p'.

DEFINITION:
p (*p*, *n*)
=
if $n \cong 0$ **then** *p*
 else p (p-step (*p*),*n* −1) **endif**

This function steps the proper p-state *p* a specified number of times, *n*.
   The single-stepper for Piton is

DEFINITION:
p-step (*p*)
=
if p-psw (*p*) = **' run** **then** p-step1 (p-current-instruction (*p*), *p*)
 else *p* **endif**

Note that 'p-step' is a no-op if the psw is not **' run**. If the psw is **' run** we use the
function 'p-step1' on the current instruction and the current p-state.

DEFINITION:
p-step1 (*ins*, *p*)
=
if p-ins-okp (*ins*, *p*) **then** p-ins-step (*ins*, *p*)
 else p-halt (*p*, x-y-error-msg (**' p**, car (*ins*))) **endif**

'p-step1' first checks the precondition of the current instruction, using 'p-ins-okp'.
If the precondition is satisfied, 'p-ins-step' is used to compute the next state. Otherwise, the psw of the current state is set to an error message.

   The two functions, 'p-ins-okp' and 'p-ins-step' are defined as case splits on the
opcode of the current instruction. For each opcode there is a function that checks the
precondition and another that computes the step. To determine the precondition of
the instruction *opcode* look up the function 'p-*opcode*-okp'. To determine the step
function for that instruction, see 'p-*opcode*-step'.

   Consider the **push-constant** instruction again. The precondition for that instruction is encoded in

DEFINITION:
p-push-constant-okp (*ins*, *p*)
=
length (p-temp-stk (*p*)) < p-max-temp-stk-size (*p*)

This function merely checks that there is room on the temporary stack to do one
more push. The syntactic constraints on proper p-states assures us that the object to
be pushed is a legal Piton object (given our treatment of the special token **pc** and of
labels). The syntactic constraints also assure us that it is legal to increment the
program counter by one (it is impossible to fall off the end of a proper Piton
program).

   The step function for **push-constant** is

DEFINITION:
p-push-constant-step (*ins*, *p*)

=

p-state (add1-p-pc (*p*),
        p-ctrl-stk (*p*),
        push (unabbreviate-constant (cadr (*ins*), *p*), p-temp-stk (*p*)),
        p-prog-segment (*p*),
        p-data-segment (*p*),
        p-max-ctrl-stk-size (*p*),
        p-max-temp-stk-size (*p*),
        p-word-size (*p*),
        '**run**)

Note that the step function increments the program counter by one, pushes one thing onto the temporary stack (obtained by "unabbreviating" the operand of the instruction), and does not alter any other component of the current state.

The precondition and effects of all other Piton instructions are defined similarly. The formal definitions below (accessed via the Index or via the 'p-*opcode*-okp'/'p-*opcode*-step' naming convention) are offered as a precise reference manual for Piton.

Readers uninterested in pursuing the formal definition at this time should skip to page 243.

II.2. Alphabetical Listing of the Piton Definitions

DEFINITION:
add-addr (*addr*, *n*) = tag (type (*addr*), add-adp (untag (*addr*), *n*))

DEFINITION:
add-adp (*adp*, *n*) = cons (adp-name (*adp*), adp-offset (*adp*) + *n*)

DEFINITION:
add1-addr (*addr*) = add-addr (*addr*, **1**)

DEFINITION:
add1-p-pc (*p*) = add1-addr (p-pc (*p*))

DEFINITION:
adp-name (*adp*) = car (*adp*)

DEFINITION:
adp-offset (*adp*) = cdr (*adp*)

DEFINITION:
adpp $(x, segment)$
=

    listp (x)
$\wedge$ (adp-offset $(x) \in \mathbf{N}$)
$\wedge$ definedp (adp-name (x), *segment*)
$\wedge$ (adp-offset (x) < length (definiens (adp-name (x), *segment*)))

DEFINITION:
all-but-last (a)
=

if nlistp (a) **then nil**
 elseif nlistp (cdr (a)) **then nil**
 else cons (car (a), all-but-last (cdr (a))) **endif**

DEFINITION:
all-find-labelp $(lab\text{-}lst, lst)$
=

if nlistp $(lab\text{-}lst)$ **then t**
 else find-labelp (car $(lab\text{-}lst)$, lst) $\wedge$ all-find-labelp (cdr $(lab\text{-}lst)$, lst) **endif**

DEFINITION:
all-litatoms (lst)
=

if nlistp (lst) **then** $lst =$ **nil**
 else litatom (car (lst)) $\wedge$ all-litatoms (cdr (lst)) **endif**

DEFINITION:
all-p-objectps (lst, p)
=

if nlistp (lst) **then** $lst =$ **nil**
 else p-objectp (car (lst), p) $\wedge$ all-p-objectps (cdr (lst), p) **endif**

DEFINITION:
all-zero-bitvp (a)
=

if listp (a) **then** (car $(a) = 0$) $\wedge$ all-zero-bitvp (cdr (a))
 else t endif

DEFINITION:
and-bit $(bit1, bit2)$
=

if $bit1 = 0$ **then 0**
 elseif $bit2 = 0$ **then 0**
 else 1 endif

DEFINITION:
and-bitv (a, b)

$=$

if nlistp (a) **then nil**
 else cons (and-bit (car (a), car (b)), and-bitv (cdr (a), cdr (b))) **endif**

DEFINITION:
and-bool (x, y)

$=$

if $x =$ ' **f** **then** ' **f**
 else y **endif**

DEFINITION:
area-name (x) = adp-name (untag (x))

DEFINITION:
bindings $(frame)$ = car $(frame)$

DEFINITION:
bit-vectorp (x, n)

$=$

if nlistp (x) **then** $(x = $ **nil**$) \wedge (n \cong 0)$
 else $(\neg (n \cong 0)) \wedge$ bitp (car (x)) $\wedge$ bit-vectorp (cdr $(x),n-1$) **endif**

DEFINITION:
bitp $(x) = (x = 0) \vee (x = 1)$

DEFINITION:
bool (x)

$=$

tag (' **bool**,
    if x **then** ' **t**
      else ' **f** **endif**)

DEFINITION:
bool-to-nat (flg)

$=$

if $flg = $ ' **f** **then** 0
 else 1 **endif**

DEFINITION:
booleanp $(x) = (x = $ ' **t**$) \vee (x = $ ' **f**$)$

DEFINITION:
definedp (*name*, *alist*)

=

if nlistp (*alist*) **then f**
 elseif *name* = caar (*alist*) **then t**
 else definedp (*name*, cdr (*alist*)) **endif**

DEFINITION:
definiens (*name*, *alist*) = cdr (definition (*name*, *alist*))

DEFINITION:
definition (*name*, *alist*) = assoc (*name*, *alist*)

DEFINITION:
deposit (*val*, *addr*, *segment*) = deposit-adp (*val*, untag (*addr*), *segment*)

DEFINITION:
deposit-adp (*val*, *adp*, *segment*)

=

put-value (put (*val*, adp-offset (*adp*), definiens (adp-name (*adp*), *segment*)),
          adp-name (*adp*),
          segment)

DEFINITION:
i^j

=

if $j \cong 0$ **then 1**
 else $i \times i^{(j-1)}$ **endif**

DEFINITION:
fall-off-proofp (*lst*)

=

   car (unlabel (get (length (*lst*) −1, *lst*)))
$\in$ `'(ret jump jump-case popj)`

DEFINITION:
fetch (*addr*, *segment*) = fetch-adp (untag (*addr*), *segment*)

DEFINITION:
fetch-adp (*adp*, *segment*)

=

get (adp-offset (*adp*), definiens (adp-name (*adp*), *segment*))

DEFINITION:
find-label (x, lst)

$=$

if nlistp (lst) **then** 0
  elseif labeledp $(car(lst)) \wedge (x = cadar(lst))$ **then** 0
  else 1+ find-label $(x, cdr(lst))$ **endif**

DEFINITION:
find-labelp (x, lst)

$=$

if nlistp (lst) **then f**
  elseif labeledp $(car(lst)) \wedge (x = cadar(lst))$ **then t**
  else find-labelp $(x, cdr(lst))$ **endif**

DEFINITION:
first-n (n, x)

$=$

if $n \cong 0$ **then nil**
  else cons $(car(x),$ first-n $(n-1, cdr(x)))$ **endif**

DEFINITION:
fix-small-integer $(i, word\text{-}size)$

$=$

if small-integerp $(i, word\text{-}size)$ **then** i
  elseif negativep (i) **then** iplus $(i, 2^{word\text{-}size})$
  else iplus $(i, -2^{word\text{-}size})$ **endif**

DEFINITION:
fix-small-natural $(n, word\text{-}size) = n$ **mod** $2^{word\text{-}size}$

DEFINITION:
formal-vars $(d) = cadr(d)$

DEFINITION:
get (n, lst)

$=$

if $n \cong 0$ **then** car (lst)
  else get $(n-1, cdr(lst))$ **endif**

DEFINITION:
idifference $(x, y) = $ iplus $(x,$ ineg $(y))$

DEFINITION:
ilessp (i, j)

=

if negativep (i)
 then if negativep (j) then negative-guts (j) < negative-guts (i)
      elseif $i = (- 0)$ then $0 < j$
      else t endif
 elseif negativep (j) then f
 else $i < j$ endif

DEFINITION:
ineg (x)

=

if negativep (x) then negative-guts (x)
 elseif $x \cong 0$ then 0
 else $- x$ endif

DEFINITION:
inegate (i)

=

if negativep (i) then negative-guts (i)
 elseif $i \cong 0$ then 0
 else $- i$ endif

DEFINITION:
integerp (x)

=

if $x \in N$ then t
 elseif negativep (x) then $\neg$ (negative-guts $(x) \cong 0$)
 else f endif

DEFINITION:
iplus (x, y)

=

if negativep (x)
 then if negativep (y)
        then if (negative-guts $(x) \cong 0$) $\wedge$ (negative-guts $(y) \cong 0$) then 0
             else $-$ (negative-guts (x) + negative-guts (y)) endif
        elseif $y <$ negative-guts (x) then $-$ (negative-guts $(x) - y$)
        else $y -$ negative-guts (x) endif
 elseif negativep (y)
 then if $x <$ negative-guts (y) then $-$ (negative-guts $(y) - x$)
      else $x -$ negative-guts (y) endif
 else $x + y$ endif

DEFINITION:
labeledp $(x) =$ car $(x) =$ 'dl

DEFINITION:
legal-labelp (*ins*) = labeledp (*ins*) $\rightarrow$ litatom (cadr (*ins*))

DEFINITION:
length (*l*)

=

if listp (*l*) **then** 1+ length (cdr (*l*))
  else 0 **endif**

DEFINITION:
local-var-value (*var*, *ctrl-stk*)

=

definiens (*var*, bindings (top (*ctrl-stk*)))

DEFINITION:
local-vars (*d*)

=

append (formal-vars (*d*), strip-cars (temp-var-dcls (*d*)))

DEFINITION:
lsh-bitv (*a*) = append (cdr (*a*), **' (0)**)

DEFINITION:
make-p-call-frame (*formal-vars*, *temp-stk*, *temp-var-dcls*, *ret-pc*)

=

p-frame (append (pair-formal-vars-with-actuals (*formal-vars*, *temp-stk*),
                pair-temps-with-initial-values (*temp-var-dcls*)),
         ret-pc)

DEFINITION:
name (*d*) = car (*d*)

DEFINITION:
not-bit (*bit*)

=

if *bit* = 0 **then** 1
  else 0 **endif**

DEFINITION:
not-bitv (*a*)

=

if nlistp (*a*) **then** **nil**
  else cons (not-bit (car (*a*)), not-bitv (cdr (*a*))) **endif**

DEFINITION:
not-bool (x)

=

if $x = $ '**f** then '**t**
 else '**f** endif

DEFINITION:
offset $(x) = $ adp-offset (untag (x))

DEFINITION:
or-bit $(bit1, bit2)$

=

if $bit1 = $ **0**
 then if $bit2 = $ **0** then **0**
      else **1** endif
 else **1** endif

DEFINITION:
or-bitv (a, b)

=

if nlistp (a) then nil
 else cons (or-bit (car (a), car (b)), or-bitv (cdr (a), cdr (b))) endif

DEFINITION:
or-bool (x, y)

=

if $x = $ '**f** then y
 else '**t** endif

DEFINITION:
p (p, n)

=

if $n \cong $ **0** then p
 else p (p-step (p), $n - 1$) endif

DEFINITION:
p-add-addr-okp (ins, p)

=

   listp (p-temp-stk (p))
 $\wedge$ listp (pop (p-temp-stk (p)))
 $\wedge$ p-objectp-type ('**nat**, top (p-temp-stk (p)), p)
 $\wedge$ p-objectp-type ('**addr**, top1 (p-temp-stk (p)), p)
 $\wedge$ p-objectp-type ('**addr**,
                     add-addr (top1 (p-temp-stk (p)), untag (top (p-temp-stk (p))))),
                  p)

DEFINITION:
p-add-addr-step (*ins, p*)
=
p-state (add1-p-pc (*p*),
       p-ctrl-stk (*p*),
       push (add-addr (top1 (p-temp-stk (*p*)), untag (top (p-temp-stk (*p*)))),
           pop (pop (p-temp-stk (*p*)))),
       p-prog-segment (*p*),
       p-data-segment (*p*),
       p-max-ctrl-stk-size (*p*),
       p-max-temp-stk-size (*p*),
       p-word-size (*p*),
       ' run)

DEFINITION:
p-add-int-okp (*ins, p*)
=
    listp (p-temp-stk (*p*))
∧ listp (pop (p-temp-stk (*p*)))
∧ p-objectp-type (**' int**, top (p-temp-stk (*p*)), *p*)
∧ p-objectp-type (**' int**, top1 (p-temp-stk (*p*)), *p*)
∧ small-integerp (iplus (untag (top1 (p-temp-stk (*p*))),
                       untag (top (p-temp-stk (*p*)))),
             p-word-size (*p*))

DEFINITION:
p-add-int-step (*ins, p*)
=
p-state (add1-p-pc (*p*),
       p-ctrl-stk (*p*),
       push (tag (**' int**,
               iplus (untag (top1 (p-temp-stk (*p*))),
                    untag (top (p-temp-stk (*p*)))))),
           pop (pop (p-temp-stk (*p*)))),
       p-prog-segment (*p*),
       p-data-segment (*p*),
       p-max-ctrl-stk-size (*p*),
       p-max-temp-stk-size (*p*),
       p-word-size (*p*),
       ' run)

DEFINITION:
p-add-int-with-carry-okp (*ins*, *p*)
=

    listp (p-temp-stk (*p*))
∧ listp (pop (p-temp-stk (*p*)))
∧ listp (pop (pop (p-temp-stk (*p*))))
∧ p-objectp-type (**'int**, top (p-temp-stk (*p*)), *p*)
∧ p-objectp-type (**'int**, top1 (p-temp-stk (*p*)), *p*)
∧ p-objectp-type (**'bool**, top2 (p-temp-stk (*p*)), *p*)

DEFINITION:
p-add-int-with-carry-step (*ins*, *p*)
=

p-state (add1-p-pc (*p*),
       p-ctrl-stk (*p*),
       push (tag (**'int**,
              fix-small-integer (
                iplus (bool-to-nat (untag (top2 (p-temp-stk (*p*)))),
                     iplus (untag (top1 (p-temp-stk (*p*))),
                         untag (top (p-temp-stk (*p*))))),
                p-word-size (*p*))),
             push (bool (¬small-integerp (
                    iplus (bool-to-nat (untag (top2 (p-temp-stk (*p*)))),
                       iplus (untag (top1 (p-temp-stk (*p*))),
                         untag (top (p-temp-stk (*p*))))),
                p-word-size (*p*))),
              pop (pop (pop (p-temp-stk (*p*))))))),
       p-prog-segment (*p*),
       p-data-segment (*p*),
       p-max-ctrl-stk-size (*p*),
       p-max-temp-stk-size (*p*),
       p-word-size (*p*),
       'run)

DEFINITION:
p-add-nat-okp (*ins*, *p*)
=

    listp (p-temp-stk (*p*))
∧ listp (pop (p-temp-stk (*p*)))
∧ p-objectp-type (**'nat**, top (p-temp-stk (*p*)), *p*)
∧ p-objectp-type (**'nat**, top1 (p-temp-stk (*p*)), *p*)
∧ small-naturalp (untag (top1 (p-temp-stk (*p*))) + untag (top (p-temp-stk (*p*))),
                 p-word-size (*p*))

DEFINITION:
p-add-nat-step (ins, p)

=

p-state (add1-p-pc (p),
        p-ctrl-stk (p),
        push (tag ('**nat**,
                   untag (top1 (p-temp-stk (p))) + untag (top (p-temp-stk (p))))),
              pop (pop (p-temp-stk (p))))),
        p-prog-segment (p),
        p-data-segment (p),
        p-max-ctrl-stk-size (p),
        p-max-temp-stk-size (p),
        p-word-size (p),
        '**run**)

DEFINITION:
p-add-nat-with-carry-okp (ins, p)

=

    listp (p-temp-stk (p))
$\wedge$ listp (pop (p-temp-stk (p)))
$\wedge$ listp (pop (pop (p-temp-stk (p))))
$\wedge$ p-objectp-type ('**nat**, top (p-temp-stk (p)), p)
$\wedge$ p-objectp-type ('**nat**, top1 (p-temp-stk (p)), p)
$\wedge$ p-objectp-type ('**bool**, top2 (p-temp-stk (p)), p)

DEFINITION:
p-add-nat-with-carry-step (ins, p)

=

p-state (add1-p-pc (p),
        p-ctrl-stk (p),
        push (tag ('**nat**,
               fix-small-natural (
                 bool-to-nat (untag (top2 (p-temp-stk (p)))))
              +   untag (top1 (p-temp-stk (p)))
              +   untag (top (p-temp-stk (p)))),
              p-word-size (p))),
           push (bool ($\neg$small-naturalp (
                   bool-to-nat (untag (top2 (p-temp-stk (p)))))
                +   untag (top1 (p-temp-stk (p)))
                +   untag (top (p-temp-stk (p)))),
                p-word-size (p))),
              pop (pop (pop (p-temp-stk (p))))))),
        p-prog-segment (p),
        p-data-segment (p),
        p-max-ctrl-stk-size (p),
        p-max-temp-stk-size (p),
        p-word-size (p),
        '**run**)

DEFINITION:
p-add1-int-okp (*ins*, *p*)
=

   listp (p-temp-stk (*p*))
∧ p-objectp-type (**'int**, top (p-temp-stk (*p*)), *p*)
∧ small-integerp (iplus (**1**, untag (top (p-temp-stk (*p*)))), p-word-size (*p*))

DEFINITION:
p-add1-int-step (*ins*, *p*)
=

p-state (add1-p-pc (*p*),
        p-ctrl-stk (*p*),
        push (tag (**'int**, iplus (**1**, untag (top (p-temp-stk (*p*))))),
             pop (p-temp-stk (*p*))),
        p-prog-segment (*p*),
        p-data-segment (*p*),
        p-max-ctrl-stk-size (*p*),
        p-max-temp-stk-size (*p*),
        p-word-size (*p*),
        'run)

DEFINITION:
p-add1-nat-okp (*ins*, *p*)
=

   listp (p-temp-stk (*p*))
∧ p-objectp-type (**'nat**, top (p-temp-stk (*p*)), *p*)
∧ small-naturalp (1+ untag (top (p-temp-stk (*p*))), p-word-size (*p*))

DEFINITION:
p-add1-nat-step (*ins*, *p*)
=

p-state (add1-p-pc (*p*),
        p-ctrl-stk (*p*),
        push (tag (**'nat**, 1+ untag (top (p-temp-stk (*p*)))), pop (p-temp-stk (*p*))),
        p-prog-segment (*p*),
        p-data-segment (*p*),
        p-max-ctrl-stk-size (*p*),
        p-max-temp-stk-size (*p*),
        p-word-size (*p*),
        'run)

DEFINITION:
p-and-bitv-okp (*ins*, *p*)
=

   listp (p-temp-stk (*p*))
∧ listp (pop (p-temp-stk (*p*)))
∧ p-objectp-type (**'bitv**, top (p-temp-stk (*p*)), *p*)
∧ p-objectp-type (**'bitv**, top1 (p-temp-stk (*p*)), *p*)

DEFINITION:
p-and-bitv-step (*ins*, *p*)
=
p-state (add1-p-pc (*p*),
      p-ctrl-stk (*p*),
      push (tag (**'bitv**,
               and-bitv (untag (top1 (p-temp-stk (*p*))),
                       untag (top (p-temp-stk (*p*))))),
           pop (pop (p-temp-stk (*p*)))),
      p-prog-segment (*p*),
      p-data-segment (*p*),
      p-max-ctrl-stk-size (*p*),
      p-max-temp-stk-size (*p*),
      p-word-size (*p*),
      'run)

DEFINITION:
p-and-bool-okp (*ins*, *p*)
=
    listp (p-temp-stk (*p*))
$\wedge$ listp (pop (p-temp-stk (*p*)))
$\wedge$ p-objectp-type (**'bool**, top (p-temp-stk (*p*)), *p*)
$\wedge$ p-objectp-type (**'bool**, top1 (p-temp-stk (*p*)), *p*)

DEFINITION:
p-and-bool-step (*ins*, *p*)
=
p-state (add1-p-pc (*p*),
      p-ctrl-stk (*p*),
      push (tag (**'bool**,
               and-bool (untag (top1 (p-temp-stk (*p*))),
                       untag (top (p-temp-stk (*p*))))),
           pop (pop (p-temp-stk (*p*)))),
      p-prog-segment (*p*),
      p-data-segment (*p*),
      p-max-ctrl-stk-size (*p*),
      p-max-temp-stk-size (*p*),
      p-word-size (*p*),
      'run)

DEFINITION:
p-call-okp (*ins*, *p*)
=
   (p-max-ctrl-stk-size (*p*)
   ≥   p-ctrl-stk-size (push (make-p-call-frame (
                                 formal-vars (definition (cadr (*ins*),
                                           p-prog-segment (*p*))),
                       p-temp-stk (*p*),
                       temp-var-dcls (definition (cadr (*ins*),
                                             p-prog-segment (*p*))),
                       add1-addr (p-pc (*p*))),
                       p-ctrl-stk (*p*))))
∧ (length (p-temp-stk (*p*))
   ≥   length (formal-vars (definition (cadr (*ins*), p-prog-segment (*p*)))))

DEFINITION:
p-call-step (*ins*, *p*)
=
p-state (tag ('**pc**, cons (cadr (*ins*), **0**)),
       push (make-p-call-frame (
            formal-vars (definition (cadr (*ins*),
                                 p-prog-segment (*p*))),
            p-temp-stk (*p*),
            temp-var-dcls (definition (cadr (*ins*),
                                     p-prog-segment (*p*))),
            add1-addr (p-pc (*p*))),
            p-ctrl-stk (*p*)),
       popn (length (formal-vars (definition (cadr (*ins*), p-prog-segment (*p*)))),
            p-temp-stk (*p*)),
       p-prog-segment (*p*),
       p-data-segment (*p*),
       p-max-ctrl-stk-size (*p*),
       p-max-temp-stk-size (*p*),
       p-word-size (*p*),
       '**run**)

DEFINITION:
p-ctrl-stk-size (*ctrl-stk*)
=
if nlistp (*ctrl-stk*) **then 0**
 else p-frame-size (top (*ctrl-stk*)) + p-ctrl-stk-size (cdr (*ctrl-stk*)) **endif**

DEFINITION:
p-current-instruction (*p*)
=
unlabel (get (offset (p-pc (*p*)), program-body (p-current-program (*p*))))

DEFINITION:
p-current-program (*p*)

=

definition (area-name (p-pc (*p*)), p-prog-segment (*p*))

DEFINITION:
p-deposit-okp (*ins*, *p*)

=

    listp (p-temp-stk (*p*))
∧ listp (pop (p-temp-stk (*p*)))
∧ p-objectp-type (**'addr**, top (p-temp-stk (*p*)), *p*)

DEFINITION:
p-deposit-step (*ins*, *p*)

=

p-state (add1-p-pc (*p*),
       p-ctrl-stk (*p*),
       pop (pop (p-temp-stk (*p*))),
       p-prog-segment (*p*),
       deposit (top1 (p-temp-stk (*p*)), top (p-temp-stk (*p*)), p-data-segment (*p*)),
       p-max-ctrl-stk-size (*p*),
       p-max-temp-stk-size (*p*),
       p-word-size (*p*),
       'run)

DEFINITION:
p-deposit-temp-stk-okp (*ins*, *p*)

=

    listp (p-temp-stk (*p*))
∧ listp (pop (p-temp-stk (*p*)))
∧ p-objectp-type (**'nat**, top (p-temp-stk (*p*)), *p*)
∧ (untag (top (p-temp-stk (*p*))) < length (pop (pop (p-temp-stk (*p*)))))

DEFINITION:
p-deposit-temp-stk-step (*ins*, *p*)

=

p-state (add1-p-pc (*p*),
       p-ctrl-stk (*p*),
       rput (top1 (p-temp-stk (*p*)),
           untag (top (p-temp-stk (*p*))),
           pop (pop (p-temp-stk (*p*)))),
       p-prog-segment (*p*),
       p-data-segment (*p*),
       p-max-ctrl-stk-size (*p*),
       p-max-temp-stk-size (*p*),
       p-word-size (*p*),
       'run)

DEFINITION:
p-div2-nat-okp (*ins*, *p*)

=

    listp (p-temp-stk (*p*))
∧ p-objectp-type (**'nat**, top (p-temp-stk (*p*)), *p*)
∧ (length (p-temp-stk (*p*)) < p-max-temp-stk-size (*p*))

DEFINITION:
p-div2-nat-step (*ins*, *p*)

=

p-state (add1-p-pc (*p*),
        p-ctrl-stk (*p*),
        push (tag (**'nat**, untag (top (p-temp-stk (*p*)))) **mod 2**),
            push (tag (**'nat**, untag (top (p-temp-stk (*p*)))) / **2**),
                pop (p-temp-stk (*p*)))),
        p-prog-segment (*p*),
        p-data-segment (*p*),
        p-max-ctrl-stk-size (*p*),
        p-max-temp-stk-size (*p*),
        p-word-size (*p*),
        'run)

DEFINITION:
p-eq-okp (*ins*, *p*)

=

    listp (p-temp-stk (*p*))
∧ listp (pop (p-temp-stk (*p*)))
∧ (type (top (p-temp-stk (*p*))) = type (top1 (p-temp-stk (*p*))))

DEFINITION:
p-eq-step (*ins*, *p*)

=

p-state (add1-p-pc (*p*),
        p-ctrl-stk (*p*),
        push (bool (untag (top1 (p-temp-stk (*p*))) = untag (top (p-temp-stk (*p*)))),
            pop (pop (p-temp-stk (*p*)))),
        p-prog-segment (*p*),
        p-data-segment (*p*),
        p-max-ctrl-stk-size (*p*),
        p-max-temp-stk-size (*p*),
        p-word-size (*p*),
        'run)

DEFINITION:
p-fetch-okp (*ins*, *p*)

=

listp (p-temp-stk (*p*)) ∧ p-objectp-type (**'addr**, top (p-temp-stk (*p*)), *p*)

DEFINITION:
p-fetch-step (*ins*, *p*)
=
p-state (add1-p-pc (*p*),
        p-ctrl-stk (*p*),
        push (fetch (top (p-temp-stk (*p*)),
                        p-data-segment (*p*)), pop (p-temp-stk (*p*))),
        p-prog-segment (*p*),
        p-data-segment (*p*),
        p-max-ctrl-stk-size (*p*),
        p-max-temp-stk-size (*p*),
        p-word-size (*p*),
        'run)

DEFINITION:
p-fetch-temp-stk-okp (*ins*, *p*)
=
    listp (p-temp-stk (*p*))
∧   p-objectp-type (**'nat**, top (p-temp-stk (*p*)), *p*)
∧   (untag (top (p-temp-stk (*p*))) < length (p-temp-stk (*p*)))

DEFINITION:
p-fetch-temp-stk-step (*ins*, *p*)
=
p-state (add1-p-pc (*p*),
        p-ctrl-stk (*p*),
        push (rget (untag (top (p-temp-stk (*p*))), p-temp-stk (*p*)),
            pop (p-temp-stk (*p*))),
        p-prog-segment (*p*),
        p-data-segment (*p*),
        p-max-ctrl-stk-size (*p*),
        p-max-temp-stk-size (*p*),
        p-word-size (*p*),
        'run)

DEFINITION:
p-frame (*bindings*, *ret-pc*) = list (*bindings*, *ret-pc*)

DEFINITION:
p-frame-size (*frame*) = **2** + length (bindings (*frame*))

DEFINITION:
p-halt (*p*, *psw*)
=
p-state (p-pc (*p*),
       p-ctrl-stk (*p*),
       p-temp-stk (*p*),
       p-prog-segment (*p*),
       p-data-segment (*p*),
       p-max-ctrl-stk-size (*p*),
       p-max-temp-stk-size (*p*),
       p-word-size (*p*),
       psw)

DEFINITION:
p-ins-okp (*ins*, *p*)
=
case on car (*ins*):
 case = **call**
 then p-call-okp (*ins*, *p*)
 case = **ret**
 then p-ret-okp (*ins*, *p*)
 case = **locn**
 then p-locn-okp (*ins*, *p*)
 case = **push-constant**
 then p-push-constant-okp (*ins*, *p*)
 case = **push-local**
 then p-push-local-okp (*ins*, *p*)
 case = **push-global**
 then p-push-global-okp (*ins*, *p*)
 case = **push-ctrl-stk-free-size**
 then p-push-ctrl-stk-free-size-okp (*ins*, *p*)
 case = **push-temp-stk-free-size**
 then p-push-temp-stk-free-size-okp (*ins*, *p*)
 case = **push-temp-stk-index**
 then p-push-temp-stk-index-okp (*ins*, *p*)
 case = **jump-if-temp-stk-full**
 then p-jump-if-temp-stk-full-okp (*ins*, *p*)
 case = **jump-if-temp-stk-empty**
 then p-jump-if-temp-stk-empty-okp (*ins*, *p*)
 case = **pop**
 then p-pop-okp (*ins*, *p*)
 case = **pop***
 then p-pop*-okp (*ins*, *p*)
 case = **popn**
 then p-popn-okp (*ins*, *p*)
 case = **pop-local**
 then p-pop-local-okp (*ins*, *p*)
 case = **pop-global**
 then p-pop-global-okp (*ins*, *p*)

case = **pop-locn**
then p-pop-locn-okp (*ins*, *p*)
case = **pop-call**
then p-pop-call-okp (*ins*, *p*)
case = **fetch-temp-stk**
then p-fetch-temp-stk-okp (*ins*, *p*)
case = **deposit-temp-stk**
then p-deposit-temp-stk-okp (*ins*, *p*)
case = **jump**
then p-jump-okp (*ins*, *p*)
case = **jump-case**
then p-jump-case-okp (*ins*, *p*)
case = **pushj**
then p-pushj-okp (*ins*, *p*)
case = **popj**
then p-popj-okp (*ins*, *p*)
case = **set-local**
then p-set-local-okp (*ins*, *p*)
case = **set-global**
then p-set-global-okp (*ins*, *p*)
case = **test-nat-and-jump**
then p-test-nat-and-jump-okp (*ins*, *p*)
case = **test-int-and-jump**
then p-test-int-and-jump-okp (*ins*, *p*)
case = **test-bool-and-jump**
then p-test-bool-and-jump-okp (*ins*, *p*)
case = **test-bitv-and-jump**
then p-test-bitv-and-jump-okp (*ins*, *p*)
case = **no-op**
then p-no-op-okp (*ins*, *p*)
case = **add-addr**
then p-add-addr-okp (*ins*, *p*)
case = **sub-addr**
then p-sub-addr-okp (*ins*, *p*)
case = **eq**
then p-eq-okp (*ins*, *p*)
case = **lt-addr**
then p-lt-addr-okp (*ins*, *p*)
case = **fetch**
then p-fetch-okp (*ins*, *p*)
case = **deposit**
then p-deposit-okp (*ins*, *p*)
case = **add-int**
then p-add-int-okp (*ins*, *p*)
case = **add-int-with-carry**
then p-add-int-with-carry-okp (*ins*, *p*)
case = **add1-int**
then p-add1-int-okp (*ins*, *p*)
case = **sub-int**
then p-sub-int-okp (*ins*, *p*)

```
case = sub-int-with-carry
then p-sub-int-with-carry-okp (ins, p)
case = sub1-int
then p-sub1-int-okp (ins, p)
case = neg-int
then p-neg-int-okp (ins, p)
case = lt-int
then p-lt-int-okp (ins, p)
case = int-to-nat
then p-int-to-nat-okp (ins, p)
case = add-nat
then p-add-nat-okp (ins, p)
case = add-nat-with-carry
then p-add-nat-with-carry-okp (ins, p)
case = add1-nat
then p-add1-nat-okp (ins, p)
case = sub-nat
then p-sub-nat-okp (ins, p)
case = sub-nat-with-carry
then p-sub-nat-with-carry-okp (ins, p)
case = sub1-nat
then p-sub1-nat-okp (ins, p)
case = lt-nat
then p-lt-nat-okp (ins, p)
case = mult2-nat
then p-mult2-nat-okp (ins, p)
case = mult2-nat-with-carry-out
then p-mult2-nat-with-carry-out-okp (ins, p)
case = div2-nat
then p-div2-nat-okp (ins, p)
case = or-bitv
then p-or-bitv-okp (ins, p)
case = and-bitv
then p-and-bitv-okp (ins, p)
case = not-bitv
then p-not-bitv-okp (ins, p)
case = xor-bitv
then p-xor-bitv-okp (ins, p)
case = rsh-bitv
then p-rsh-bitv-okp (ins, p)
case = lsh-bitv
then p-lsh-bitv-okp (ins, p)
case = or-bool
then p-or-bool-okp (ins, p)
case = and-bool
then p-and-bool-okp (ins, p)
case = not-bool
then p-not-bool-okp (ins, p)
otherwise f  endcase
```

DEFINITION:
p-ins-step (*ins*, *p*)
=
case on car (*ins*):
 case = **call**
 then p-call-step (*ins*, *p*)
 case = **ret**
 then p-ret-step (*ins*, *p*)
 case = **locn**
 then p-locn-step (*ins*, *p*)
 case = **push-constant**
 then p-push-constant-step (*ins*, *p*)
 case = **push-local**
 then p-push-local-step (*ins*, *p*)
 case = **push-global**
 then p-push-global-step (*ins*, *p*)
 case = **push-ctrl-stk-free-size**
 then p-push-ctrl-stk-free-size-step (*ins*, *p*)
 case = **push-temp-stk-free-size**
 then p-push-temp-stk-free-size-step (*ins*, *p*)
 case = **push-temp-stk-index**
 then p-push-temp-stk-index-step (*ins*, *p*)
 case = **jump-if-temp-stk-full**
 then p-jump-if-temp-stk-full-step (*ins*, *p*)
 case = **jump-if-temp-stk-empty**
 then p-jump-if-temp-stk-empty-step (*ins*, *p*)
 case = **pop**
 then p-pop-step (*ins*, *p*)
 case = **pop***
 then p-pop*-step (*ins*, *p*)
 case = **popn**
 then p-popn-step (*ins*, *p*)
 case = **pop-local**
 then p-pop-local-step (*ins*, *p*)
 case = **pop-global**
 then p-pop-global-step (*ins*, *p*)
 case = **pop-locn**
 then p-pop-locn-step (*ins*, *p*)
 case = **pop-call**
 then p-pop-call-step (*ins*, *p*)
 case = **fetch-temp-stk**
 then p-fetch-temp-stk-step (*ins*, *p*)
 case = **deposit-temp-stk**
 then p-deposit-temp-stk-step (*ins*, *p*)
 case = **jump**
 then p-jump-step (*ins*, *p*)
 case = **jump-case**
 then p-jump-case-step (*ins*, *p*)
 case = **pushj**
 then p-pushj-step (*ins*, *p*)

case = **popj**
then p-popj-step (ins, p)
case = **set-local**
then p-set-local-step (ins, p)
case = **set-global**
then p-set-global-step (ins, p)
case = **test-nat-and-jump**
then p-test-nat-and-jump-step (ins, p)
case = **test-int-and-jump**
then p-test-int-and-jump-step (ins, p)
case = **test-bool-and-jump**
then p-test-bool-and-jump-step (ins, p)
case = **test-bitv-and-jump**
then p-test-bitv-and-jump-step (ins, p)
case = **no-op**
then p-no-op-step (ins, p)
case = **add-addr**
then p-add-addr-step (ins, p)
case = **sub-addr**
then p-sub-addr-step (ins, p)
case = **eq**
then p-eq-step (ins, p)
case = **lt-addr**
then p-lt-addr-step (ins, p)
case = **fetch**
then p-fetch-step (ins, p)
case = **deposit**
then p-deposit-step (ins, p)
case = **add-int**
then p-add-int-step (ins, p)
case = **add-int-with-carry**
then p-add-int-with-carry-step (ins, p)
case = **add1-int**
then p-add1-int-step (ins, p)
case = **sub-int**
then p-sub-int-step (ins, p)
case = **sub-int-with-carry**
then p-sub-int-with-carry-step (ins, p)
case = **sub1-int**
then p-sub1-int-step (ins, p)
case = **neg-int**
then p-neg-int-step (ins, p)
case = **lt-int**
then p-lt-int-step (ins, p)
case = **int-to-nat**
then p-int-to-nat-step (ins, p)
case = **add-nat**
then p-add-nat-step (ins, p)
case = **add-nat-with-carry**
then p-add-nat-with-carry-step (ins, p)

case = **add1-nat**
then p-add1-nat-step (*ins*, *p*)
case = **sub-nat**
then p-sub-nat-step (*ins*, *p*)
case = **sub-nat-with-carry**
then p-sub-nat-with-carry-step (*ins*, *p*)
case = **sub1-nat**
then p-sub1-nat-step (*ins*, *p*)
case = **lt-nat**
then p-lt-nat-step (*ins*, *p*)
case = **mult2-nat**
then p-mult2-nat-step (*ins*, *p*)
case = **mult2-nat-with-carry-out**
then p-mult2-nat-with-carry-out-step (*ins*, *p*)
case = **div2-nat**
then p-div2-nat-step (*ins*, *p*)
case = **or-bitv**
then p-or-bitv-step (*ins*, *p*)
case = **and-bitv**
then p-and-bitv-step (*ins*, *p*)
case = **not-bitv**
then p-not-bitv-step (*ins*, *p*)
case = **xor-bitv**
then p-xor-bitv-step (*ins*, *p*)
case = **rsh-bitv**
then p-rsh-bitv-step (*ins*, *p*)
case = **lsh-bitv**
then p-lsh-bitv-step (*ins*, *p*)
case = **or-bool**
then p-or-bool-step (*ins*, *p*)
case = **and-bool**
then p-and-bool-step (*ins*, *p*)
case = **not-bool**
then p-not-bool-step (*ins*, *p*)
otherwise p-halt (*p*, **'run**)   **endcase**

DEFINITION:
p-int-to-nat-okp (*ins*, *p*)
=
    listp (p-temp-stk (*p*))
$\wedge$ p-objectp-type (**'int**, top (p-temp-stk (*p*)), *p*)
$\wedge$ ($\neg$ negativep (untag (top (p-temp-stk (*p*)))))

DEFINITION:
p-int-to-nat-step *(ins, p)*

=

p-state (add1-p-pc *(p)*,
       p-ctrl-stk *(p)*,
       push (tag (**'nat**, untag (top (p-temp-stk *(p)*)))), pop (p-temp-stk *(p)*))),
       p-prog-segment *(p)*,
       p-data-segment *(p)*,
       p-max-ctrl-stk-size *(p)*,
       p-max-temp-stk-size *(p)*,
       p-word-size *(p)*,
       'run)

DEFINITION:
p-jump-case-okp *(ins, p)*

=

    listp (p-temp-stk *(p)*)
∧ p-objectp-type (**'nat**, top (p-temp-stk *(p)*)), *p*)
∧ (untag (top (p-temp-stk *(p)*))) < length (cdr *(ins)*))

DEFINITION:
p-jump-case-step *(ins, p)*

=

p-state (pc (get (untag (top (p-temp-stk *(p)*))), cdr *(ins)*), p-current-program *(p)*),
       p-ctrl-stk *(p)*,
       pop (p-temp-stk *(p)*),
       p-prog-segment *(p)*,
       p-data-segment *(p)*,
       p-max-ctrl-stk-size *(p)*,
       p-max-temp-stk-size *(p)*,
       p-word-size *(p)*,
       'run)

DEFINITION:
p-jump-if-temp-stk-empty-okp *(ins, p)* = **t**

DEFINITION:
p-jump-if-temp-stk-empty-step (*ins*, *p*)
=
p-state (**if** length (p-temp-stk (*p*)) ≅ 0
   then pc (cadr (*ins*), p-current-program (*p*))
   else add1-p-pc (*p*) **endif**,
   p-ctrl-stk (*p*),
   p-temp-stk (*p*),
   p-prog-segment (*p*),
   p-data-segment (*p*),
   p-max-ctrl-stk-size (*p*),
   p-max-temp-stk-size (*p*),
   p-word-size (*p*),
   ' run)

DEFINITION:
p-jump-if-temp-stk-full-okp (*ins*, *p*) = **t**

DEFINITION:
p-jump-if-temp-stk-full-step (*ins*, *p*)
=
p-state (**if** length (p-temp-stk (*p*)) = p-max-temp-stk-size (*p*)
   then pc (cadr (*ins*), p-current-program (*p*))
   else add1-p-pc (*p*) **endif**,
   p-ctrl-stk (*p*),
   p-temp-stk (*p*),
   p-prog-segment (*p*),
   p-data-segment (*p*),
   p-max-ctrl-stk-size (*p*),
   p-max-temp-stk-size (*p*),
   p-word-size (*p*),
   ' run)

DEFINITION:
p-jump-okp (*ins*, *p*) = **t**

DEFINITION:
p-jump-step (*ins*, *p*)
=
p-state (pc (cadr (*ins*), p-current-program (*p*)),
   p-ctrl-stk (*p*),
   p-temp-stk (*p*),
   p-prog-segment (*p*),
   p-data-segment (*p*),
   p-max-ctrl-stk-size (*p*),
   p-max-temp-stk-size (*p*),
   p-word-size (*p*),
   ' run)

DEFINITION:
p-locn-okp (*ins*, *p*)
=

    p-objectp-type ('**nat**, local-var-value (cadr (*ins*), p-ctrl-stk (*p*)), *p*)
∧ (untag (local-var-value (cadr (*ins*), p-ctrl-stk (*p*)))
   < length (bindings (top (p-ctrl-stk (*p*)))))
∧ (length (p-temp-stk (*p*)) < p-max-temp-stk-size (*p*))

DEFINITION:
p-locn-step (*ins*, *p*)
=

p-state (add1-p-pc (*p*),
      p-ctrl-stk (*p*),
      push (cdr (get (untag (local-var-value (cadr (*ins*), p-ctrl-stk (*p*))),
                 bindings (top (p-ctrl-stk (*p*))))),
          p-temp-stk (*p*)),
      p-prog-segment (*p*),
      p-data-segment (*p*),
      p-max-ctrl-stk-size (*p*),
      p-max-temp-stk-size (*p*),
      p-word-size (*p*),
      '**run**)

DEFINITION:
p-lsh-bitv-okp (*ins*, *p*)
=

listp (p-temp-stk (*p*)) ∧ p-objectp-type ('**bitv**, top (p-temp-stk (*p*)), *p*)

DEFINITION:
p-lsh-bitv-step (*ins*, *p*)
=

p-state (add1-p-pc (*p*),
      p-ctrl-stk (*p*),
      push (tag ('**bitv**, lsh-bitv (untag (top (p-temp-stk (*p*))))),
         pop (p-temp-stk (*p*))),
      p-prog-segment (*p*),
      p-data-segment (*p*),
      p-max-ctrl-stk-size (*p*),
      p-max-temp-stk-size (*p*),
      p-word-size (*p*),
      '**run**)

DEFINITION:
p-lt-addr-okp (*ins*, *p*)

=

    listp (p-temp-stk (*p*))

∧ listp (pop (p-temp-stk (*p*)))

∧ p-objectp-type (**'addr**, top (p-temp-stk (*p*)), *p*)

∧ p-objectp-type (**'addr**, top1 (p-temp-stk (*p*)), *p*)

∧ (area-name (top (p-temp-stk (*p*))) = area-name (top1 (p-temp-stk (*p*))))

DEFINITION:
p-lt-addr-step (*ins*, *p*)

=

p-state (add1-p-pc (*p*),
      p-ctrl-stk (*p*),
      push (bool (offset (top1 (p-temp-stk (*p*)))
               < offset (top (p-temp-stk (*p*))))),
          pop (pop (p-temp-stk (*p*)))),
      p-prog-segment (*p*),
      p-data-segment (*p*),
      p-max-ctrl-stk-size (*p*),
      p-max-temp-stk-size (*p*),
      p-word-size (*p*),
      'run)

DEFINITION:
p-lt-int-okp (*ins*, *p*)

=

    listp (p-temp-stk (*p*))

∧ listp (pop (p-temp-stk (*p*)))

∧ p-objectp-type (**'int**, top (p-temp-stk (*p*)), *p*)

∧ p-objectp-type (**'int**, top1 (p-temp-stk (*p*)), *p*)

DEFINITION:
p-lt-int-step (*ins*, *p*)

=

p-state (add1-p-pc (*p*),
      p-ctrl-stk (*p*),
      push (bool (ilessp (untag (top1 (p-temp-stk (*p*))),
                  untag (top (p-temp-stk (*p*))))),
          pop (pop (p-temp-stk (*p*)))),
      p-prog-segment (*p*),
      p-data-segment (*p*),
      p-max-ctrl-stk-size (*p*),
      p-max-temp-stk-size (*p*),
      p-word-size (*p*),
      'run)

DEFINITION:
p-lt-nat-okp (*ins, p*)
=

    listp (p-temp-stk (*p*))
∧ listp (pop (p-temp-stk (*p*)))
∧ p-objectp-type (**'nat**, top (p-temp-stk (*p*)), *p*)
∧ p-objectp-type (**'nat**, top1 (p-temp-stk (*p*)), *p*)

DEFINITION:
p-lt-nat-step (*ins, p*)
=

p-state (add1-p-pc (*p*),
       p-ctrl-stk (*p*),
       push (bool (untag (top1 (p-temp-stk (*p*))) < untag (top (p-temp-stk (*p*)))),
            pop (pop (p-temp-stk (*p*)))),
       p-prog-segment (*p*),
       p-data-segment (*p*),
       p-max-ctrl-stk-size (*p*),
       p-max-temp-stk-size (*p*),
       p-word-size (*p*),
       'run)

DEFINITION:
p-mult2-nat-okp (*ins, p*)
=

    listp (p-temp-stk (*p*))
∧ p-objectp-type (**'nat**, top (p-temp-stk (*p*)), *p*)
∧ small-naturalp (**2** × untag (top (p-temp-stk (*p*))), p-word-size (*p*))

DEFINITION:
p-mult2-nat-step (*ins, p*)
=

p-state (add1-p-pc (*p*),
       p-ctrl-stk (*p*),
       push (tag (**'nat**, **2** × untag (top (p-temp-stk (*p*))))),
            pop (p-temp-stk (*p*))),
       p-prog-segment (*p*),
       p-data-segment (*p*),
       p-max-ctrl-stk-size (*p*),
       p-max-temp-stk-size (*p*),
       p-word-size (*p*),
       'run)

DEFINITION:
p-mult2-nat-with-carry-out-okp (*ins*, *p*)
=

    listp (p-temp-stk (*p*))
∧ p-objectp-type (**'nat**, top (p-temp-stk (*p*)), *p*)
∧ (length (p-temp-stk (*p*)) < p-max-temp-stk-size (*p*))

DEFINITION:
p-mult2-nat-with-carry-out-step (*ins*, *p*)
=

p-state (add1-p-pc (*p*),
       p-ctrl-stk (*p*),
       push (tag (**'nat**,
               fix-small-natural (**2** × untag (top (p-temp-stk (*p*))),
                             p-word-size (*p*))),
             push (bool (¬ small-naturalp (**2** × untag (top (p-temp-stk (*p*))),
                                p-word-size (*p*))),
                 pop (p-temp-stk (*p*))))),
       p-prog-segment (*p*),
       p-data-segment (*p*),
       p-max-ctrl-stk-size (*p*),
       p-max-temp-stk-size (*p*),
       p-word-size (*p*),
       'run)

DEFINITION:
p-neg-int-okp (*ins*, *p*)
=

    listp (p-temp-stk (*p*))
∧ p-objectp-type (**'int**, top (p-temp-stk (*p*)), *p*)
∧ small-integerp (inegate (untag (top (p-temp-stk (*p*)))), p-word-size (*p*))

DEFINITION:
p-neg-int-step (*ins*, *p*)
=

p-state (add1-p-pc (*p*),
      p-ctrl-stk (*p*),
      push (tag (**'int**, inegate (untag (top (p-temp-stk (*p*))))),
          pop (p-temp-stk (*p*))),
      p-prog-segment (*p*),
      p-data-segment (*p*),
      p-max-ctrl-stk-size (*p*),
      p-max-temp-stk-size (*p*),
      p-word-size (*p*),
      'run)

DEFINITION:
p-no-op-okp (*ins*, *p*) = **t**

DEFINITION:
p-no-op-step (*ins*, *p*)
=
p-state (add1-p-pc (*p*),
       p-ctrl-stk (*p*),
       p-temp-stk (*p*),
       p-prog-segment (*p*),
       p-data-segment (*p*),
       p-max-ctrl-stk-size (*p*),
       p-max-temp-stk-size (*p*),
       p-word-size (*p*),
       `'run`)

DEFINITION:
p-not-bitv-okp (*ins*, *p*)
=
listp (p-temp-stk (*p*)) $\land$ p-objectp-type (`'bitv`, top (p-temp-stk (*p*)), *p*)

DEFINITION:
p-not-bitv-step (*ins*, *p*)
=
p-state (add1-p-pc (*p*),
       p-ctrl-stk (*p*),
       push (tag (`'bitv`, not-bitv (untag (top (p-temp-stk (*p*)))))),
           pop (p-temp-stk (*p*))),
       p-prog-segment (*p*),
       p-data-segment (*p*),
       p-max-ctrl-stk-size (*p*),
       p-max-temp-stk-size (*p*),
       p-word-size (*p*),
       `'run`)

DEFINITION:
p-not-bool-okp (*ins*, *p*)
=
listp (p-temp-stk (*p*)) $\land$ p-objectp-type (`'bool`, top (p-temp-stk (*p*)), *p*)

DEFINITION:
p-not-bool-step *(ins, p)*

=

p-state (add1-p-pc *(p)*,
      p-ctrl-stk *(p)*,
      push (tag (**'bool**, not-bool (untag (top (p-temp-stk *(p)*))))),
          pop (p-temp-stk *(p)*))),
      p-prog-segment *(p)*,
      p-data-segment *(p)*,
      p-max-ctrl-stk-size *(p)*,
      p-max-temp-stk-size *(p)*,
      p-word-size *(p)*,
      'run)

DEFINITION:
p-objectp *(x, p)*

=

   listp *(x)*
$\wedge$ (cddr *(x)* = **nil**)
$\wedge$ **case on** type *(x)*:
     case = nat then small-naturalp (untag *(x)*, p-word-size *(p)*)
     case = int then small-integerp (untag *(x)*, p-word-size *(p)*)
     case = bitv then bit-vectorp (untag *(x)*, p-word-size *(p)*)
     case = bool then booleanp (untag *(x)*)
     case = addr then adpp (untag *(x)*, p-data-segment *(p)*)
     case = pc then pcpp (untag *(x)*, p-prog-segment *(p)*)
     case = subr then definedp (untag *(x)*, p-prog-segment *(p)*)
    otherwise f endcase

DEFINITION:
p-objectp-type *(type, x, p)* = (type *(x)* = *type*) $\wedge$ p-objectp *(x, p)*

DEFINITION:
p-or-bitv-okp *(ins, p)*

=

   listp (p-temp-stk *(p)*)
$\wedge$ listp (pop (p-temp-stk *(p)*))
$\wedge$ p-objectp-type (**'bitv**, top (p-temp-stk *(p)*), *p*)
$\wedge$ p-objectp-type (**'bitv**, top1 (p-temp-stk *(p)*), *p*)

DEFINITION:
p-or-bitv-step (*ins*, *p*)
=
p-state (add1-p-pc (*p*),
      p-ctrl-stk (*p*),
      push (tag (**'bitv**,
                 or-bitv (untag (top1 (p-temp-stk (*p*))),
                      untag (top (p-temp-stk (*p*))))),
            pop (pop (p-temp-stk (*p*)))),
      p-prog-segment (*p*),
      p-data-segment (*p*),
      p-max-ctrl-stk-size (*p*),
      p-max-temp-stk-size (*p*),
      p-word-size (*p*),
      'run)

DEFINITION:
p-or-bool-okp (*ins*, *p*)
=
    listp (p-temp-stk (*p*))
$\wedge$ listp (pop (p-temp-stk (*p*)))
$\wedge$ p-objectp-type (**'bool**, top (p-temp-stk (*p*)), *p*)
$\wedge$ p-objectp-type (**'bool**, top1 (p-temp-stk (*p*)), *p*)

DEFINITION:
p-or-bool-step (*ins*, *p*)
=
p-state (add1-p-pc (*p*),
      p-ctrl-stk (*p*),
      push (tag (**'bool**,
                 or-bool (untag (top1 (p-temp-stk (*p*))),
                      untag (top (p-temp-stk (*p*))))),
            pop (pop (p-temp-stk (*p*)))),
      p-prog-segment (*p*),
      p-data-segment (*p*),
      p-max-ctrl-stk-size (*p*),
      p-max-temp-stk-size (*p*),
      p-word-size (*p*),
      'run)

DEFINITION:
p-pop*-okp (*ins*, *p*) = length (p-temp-stk (*p*)) $\geq$ cadr (*ins*)

DEFINITION:
p-pop*-step (*ins*, *p*)
=

p-state (add1-p-pc (*p*),
        p-ctrl-stk (*p*),
        popn (cadr (*ins*), p-temp-stk (*p*)),
        p-prog-segment (*p*),
        p-data-segment (*p*),
        p-max-ctrl-stk-size (*p*),
        p-max-temp-stk-size (*p*),
        p-word-size (*p*),
        'run)

DEFINITION:
p-pop-call-okp (*ins*, *p*)
=

    listp (p-temp-stk (*p*))
∧   p-objectp-type (**'subr**, top (p-temp-stk (*p*)), *p*)
∧   p-call-okp (list (**'call**, untag (top (p-temp-stk (*p*)))),
                    p-state (p-pc (*p*),
                            p-ctrl-stk (*p*),
                            pop (p-temp-stk (*p*)),
                            p-prog-segment (*p*),
                            p-data-segment (*p*),
                            p-max-ctrl-stk-size (*p*),
                            p-max-temp-stk-size (*p*),
                            p-word-size (*p*),
                            'run))

DEFINITION:
p-pop-call-step (*ins*, *p*)
=

p-call-step (list (**'call**, untag (top (p-temp-stk (*p*)))),
            p-state (p-pc (*p*),
                    p-ctrl-stk (*p*),
                    pop (p-temp-stk (*p*)),
                    p-prog-segment (*p*),
                    p-data-segment (*p*),
                    p-max-ctrl-stk-size (*p*),
                    p-max-temp-stk-size (*p*),
                    p-word-size (*p*),
                    'run))

DEFINITION:
p-pop-global-okp (*ins*, *p*) = listp (p-temp-stk (*p*))

DEFINITION:
p-pop-global-step (*ins*, *p*)
=
p-state (add1-p-pc (*p*),
      p-ctrl-stk (*p*),
      pop (p-temp-stk (*p*)),
      p-prog-segment (*p*),
      deposit (top (p-temp-stk (*p*)),
              tag ('**addr**, cons (cadr (*ins*), **0**)),
              p-data-segment (*p*)),
      p-max-ctrl-stk-size (*p*),
      p-max-temp-stk-size (*p*),
      p-word-size (*p*),
      '**run**)

DEFINITION:
p-pop-local-okp (*ins*, *p*) = listp (p-temp-stk (*p*))

DEFINITION:
p-pop-local-step (*ins*, *p*)
=
p-state (add1-p-pc (*p*),
      set-local-var-value (top (p-temp-stk (*p*)), cadr (*ins*), p-ctrl-stk (*p*)),
      pop (p-temp-stk (*p*)),
      p-prog-segment (*p*),
      p-data-segment (*p*),
      p-max-ctrl-stk-size (*p*),
      p-max-temp-stk-size (*p*),
      p-word-size (*p*),
      '**run**)

DEFINITION:
p-pop-locn-okp (*ins*, *p*)
=
    p-objectp-type ('**nat**, local-var-value (cadr (*ins*), p-ctrl-stk (*p*)), *p*)
$\wedge$ (untag (local-var-value (cadr (*ins*), p-ctrl-stk (*p*)))
    < length (bindings (top (p-ctrl-stk (*p*)))))
$\wedge$ listp (p-temp-stk (*p*))

DEFINITION:
p-pop-locn-step *(ins, p)*

=

p-state (add1-p-pc *(p)*,
      set-local-var-indirect (top (p-temp-stk *(p)*),
                            untag (local-var-value (cadr *(ins)*, p-ctrl-stk *(p)*))),
                            p-ctrl-stk *(p)*),
      pop (p-temp-stk *(p)*),
      p-prog-segment *(p)*,
      p-data-segment *(p)*,
      p-max-ctrl-stk-size *(p)*,
      p-max-temp-stk-size *(p)*,
      p-word-size *(p)*,
      ' run)

DEFINITION:
p-pop-okp *(ins, p)* = listp (p-temp-stk *(p)*)

DEFINITION:
p-pop-step *(ins, p)*

=

p-state (add1-p-pc *(p)*,
      p-ctrl-stk *(p)*,
      pop (p-temp-stk *(p)*),
      p-prog-segment *(p)*,
      p-data-segment *(p)*,
      p-max-ctrl-stk-size *(p)*,
      p-max-temp-stk-size *(p)*,
      p-word-size *(p)*,
      ' run)

DEFINITION:
p-popj-okp *(ins, p)*

=

    listp (p-temp-stk *(p)*)
∧ p-objectp-type (**' pc**, top (p-temp-stk *(p)*), *p*)
∧ (area-name (top (p-temp-stk *(p)*))) = area-name (p-pc *(p)*))

DEFINITION:
p-popj-step (*ins*, *p*)
=
p-state (top (p-temp-stk (*p*)),
        p-ctrl-stk (*p*),
        pop (p-temp-stk (*p*)),
        p-prog-segment (*p*),
        p-data-segment (*p*),
        p-max-ctrl-stk-size (*p*),
        p-max-temp-stk-size (*p*),
        p-word-size (*p*),
        'run)

DEFINITION:
p-popn-okp (*ins*, *p*)
=
    listp (p-temp-stk (*p*))
∧ p-objectp-type (**'nat**, top (p-temp-stk (*p*)), *p*)
∧ (length (p-temp-stk (*p*)) ≥ (1+ untag (top (p-temp-stk (*p*)))))

DEFINITION:
p-popn-step (*ins*, *p*)
=
p-state (add1-p-pc (*p*),
        p-ctrl-stk (*p*),
        popn (untag (top (p-temp-stk (*p*))), pop (p-temp-stk (*p*))),
        p-prog-segment (*p*),
        p-data-segment (*p*),
        p-max-ctrl-stk-size (*p*),
        p-max-temp-stk-size (*p*),
        p-word-size (*p*),
        'run)

DEFINITION:
p-push-constant-okp (*ins*, *p*)
=
length (p-temp-stk (*p*)) < p-max-temp-stk-size (*p*)

DEFINITION:
p-push-constant-step (*ins*, *p*)
=

p-state (add1-p-pc (*p*),
      p-ctrl-stk (*p*),
      push (unabbreviate-constant (cadr (*ins*), *p*), p-temp-stk (*p*)),
      p-prog-segment (*p*),
      p-data-segment (*p*),
      p-max-ctrl-stk-size (*p*),
      p-max-temp-stk-size (*p*),
      p-word-size (*p*),
      'run)

DEFINITION:
p-push-ctrl-stk-free-size-okp (*ins*, *p*)
=

length (p-temp-stk (*p*)) < p-max-temp-stk-size (*p*)

DEFINITION:
p-push-ctrl-stk-free-size-step (*ins*, *p*)
=

p-state (add1-p-pc (*p*),
      p-ctrl-stk (*p*),
      push (tag (**'nat**,
                  p-max-ctrl-stk-size (*p*) − p-ctrl-stk-size (p-ctrl-stk (*p*))),
         p-temp-stk (*p*)),
      p-prog-segment (*p*),
      p-data-segment (*p*),
      p-max-ctrl-stk-size (*p*),
      p-max-temp-stk-size (*p*),
      p-word-size (*p*),
      'run)

DEFINITION:
p-push-global-okp (*ins*, *p*)
=

length (p-temp-stk (*p*)) < p-max-temp-stk-size (*p*)

DEFINITION:
p-push-global-step *(ins, p)*
=
p-state (add1-p-pc *(p)*,
      p-ctrl-stk *(p)*,
      push (fetch (tag (*' addr*, cons (cadr *(ins)*, **0**)), p-data-segment *(p)*),
          p-temp-stk *(p)*),
      p-prog-segment *(p)*,
      p-data-segment *(p)*,
      p-max-ctrl-stk-size *(p)*,
      p-max-temp-stk-size *(p)*,
      p-word-size *(p)*,
      ' run)

DEFINITION:
p-push-local-okp *(ins, p)*
=
length (p-temp-stk *(p)*) < p-max-temp-stk-size *(p)*

DEFINITION:
p-push-local-step *(ins, p)*
=
p-state (add1-p-pc *(p)*,
      p-ctrl-stk *(p)*,
      push (local-var-value (cadr *(ins)*, p-ctrl-stk *(p)*), p-temp-stk *(p)*),
      p-prog-segment *(p)*,
      p-data-segment *(p)*,
      p-max-ctrl-stk-size *(p)*,
      p-max-temp-stk-size *(p)*,
      p-word-size *(p)*,
      ' run)

DEFINITION:
p-push-temp-stk-free-size-okp *(ins, p)*
=
length (p-temp-stk *(p)*) < p-max-temp-stk-size *(p)*

DEFINITION:
p-push-temp-stk-free-size-step (*ins*, *p*)

=

p-state (add1-p-pc (*p*),
      p-ctrl-stk (*p*),
      push (tag (**'nat**, p-max-temp-stk-size (*p*) − length (p-temp-stk (*p*))),
          p-temp-stk (*p*)),
      p-prog-segment (*p*),
      p-data-segment (*p*),
      p-max-ctrl-stk-size (*p*),
      p-max-temp-stk-size (*p*),
      p-word-size (*p*),
      'run)

DEFINITION:
p-push-temp-stk-index-okp (*ins*, *p*)

=

    (length (p-temp-stk (*p*)) < p-max-temp-stk-size (*p*))
∧ (cadr (*ins*) < length (p-temp-stk (*p*)))

DEFINITION:
p-push-temp-stk-index-step (*ins*, *p*)

=

p-state (add1-p-pc (*p*),
      p-ctrl-stk (*p*),
      push (tag (**'nat**,(length (p-temp-stk (*p*)) − cadr (*ins*)) −1),
          p-temp-stk (*p*)),
      p-prog-segment (*p*),
      p-data-segment (*p*),
      p-max-ctrl-stk-size (*p*),
      p-max-temp-stk-size (*p*),
      p-word-size (*p*),
      'run)

DEFINITION:
p-pushj-okp (*ins*, *p*)

=

length (p-temp-stk (*p*)) < p-max-temp-stk-size (*p*)

DEFINITION:
p-pushj-step *(ins, p)*

=

p-state (pc (cadr *(ins)*, p-current-program *(p)*),
      p-ctrl-stk *(p)*,
      push (add1-p-pc *(p)*, p-temp-stk *(p)*),
      p-prog-segment *(p)*,
      p-data-segment *(p)*,
      p-max-ctrl-stk-size *(p)*,
      p-max-temp-stk-size *(p)*,
      p-word-size *(p)*,
      'run)

DEFINITION:
p-ret-okp *(ins, p)* = **t**

DEFINITION:
p-ret-step *(ins, p)*

=

if listp (pop (p-ctrl-stk *(p)*))
 then p-state (ret-pc (top (p-ctrl-stk *(p)*))),
          pop (p-ctrl-stk *(p)*),
          p-temp-stk *(p)*,
          p-prog-segment *(p)*,
          p-data-segment *(p)*,
          p-max-ctrl-stk-size *(p)*,
          p-max-temp-stk-size *(p)*,
          p-word-size *(p)*,
          'run)
 else p-halt *(p*, **'halt**) **endif**

DEFINITION:
p-rsh-bitv-okp *(ins, p)*

=

listp (p-temp-stk *(p)*) $\wedge$ p-objectp-type (**'bitv**, top (p-temp-stk *(p)*), *p*)

DEFINITION:
p-rsh-bitv-step *(ins, p)*
=
p-state (add1-p-pc *(p)*,
       p-ctrl-stk *(p)*,
       push (tag (**'bitv**, rsh-bitv (untag (top (p-temp-stk *(p)*))))),
           pop (p-temp-stk *(p)*))),
       p-prog-segment *(p)*,
       p-data-segment *(p)*,
       p-max-ctrl-stk-size *(p)*,
       p-max-temp-stk-size *(p)*,
       p-word-size *(p)*,
       'run)

DEFINITION:
p-set-global-okp *(ins, p)* = listp (p-temp-stk *(p)*)

DEFINITION:
p-set-global-step *(ins, p)*
=
p-state (add1-p-pc *(p)*,
       p-ctrl-stk *(p)*,
       p-temp-stk *(p)*,
       p-prog-segment *(p)*,
       deposit (top (p-temp-stk *(p)*),
              tag (**'addr**, cons (cadr *(ins)*, **0**)),
              p-data-segment *(p)*)),
       p-max-ctrl-stk-size *(p)*,
       p-max-temp-stk-size *(p)*,
       p-word-size *(p)*,
       'run)

DEFINITION:
p-set-local-okp *(ins, p)* = listp (p-temp-stk *(p)*)

DEFINITION:
p-set-local-step *(ins, p)*
=
p-state (add1-p-pc *(p)*,
       set-local-var-value (top (p-temp-stk *(p)*), cadr *(ins)*, p-ctrl-stk *(p)*),
       p-temp-stk *(p)*,
       p-prog-segment *(p)*,
       p-data-segment *(p)*,
       p-max-ctrl-stk-size *(p)*,
       p-max-temp-stk-size *(p)*,
       p-word-size *(p)*,
       'run)

SHELL DEFINITION:
Add the shell 'p-state' of 9 arguments, with
recognizer function symbol 'p-statep', and
accessors 'p-pc',
          'p-ctrl-stk', 'p-temp-stk',
          'p-prog-segment', 'p-data-segment',
          'p-max-ctrl-stk-size', 'p-max-temp-stk-size',
          'p-word-size', and 'p-psw'.

DEFINITION:
p-step (p)
=
if p-psw (p) = **'run** then p-step1 (p-current-instruction (p), p)
 else p **endif**

DEFINITION:
p-step1 (ins, p)
=
if p-ins-okp (ins, p) then p-ins-step (ins, p)
 else p-halt $(p,$ x-y-error-msg (**'p**, car $(ins)))$ **endif**

DEFINITION:
p-sub-addr-okp (ins, p)
=
   listp (p-temp-stk (p))
∧ listp (pop (p-temp-stk (p)))
∧ p-objectp-type (**'nat**, top (p-temp-stk (p)), p)
∧ p-objectp-type (**'addr**, top1 (p-temp-stk (p)), p)
∧ (offset (top1 (p-temp-stk (p))) ≥ untag (top (p-temp-stk (p))))

DEFINITION:
p-sub-addr-step (ins, p)
=
p-state (add1-p-pc (p),
       p-ctrl-stk (p),
       push (sub-addr (top1 (p-temp-stk (p)), untag (top (p-temp-stk (p))))),
          pop (pop (p-temp-stk (p))))),
       p-prog-segment (p),
       p-data-segment (p),
       p-max-ctrl-stk-size (p),
       p-max-temp-stk-size (p),
       p-word-size (p),
       'run)

DEFINITION:
p-sub-int-okp (*ins*, *p*)

=

    listp (p-temp-stk (*p*))
∧ listp (pop (p-temp-stk (*p*)))
∧ p-objectp-type ('**int**, top (p-temp-stk (*p*)), *p*)
∧ p-objectp-type ('**int**, top1 (p-temp-stk (*p*)), *p*)
∧ small-integerp (idifference (untag (top1 (p-temp-stk (*p*))),
                                    untag (top (p-temp-stk (*p*)))),
                 p-word-size (*p*))

DEFINITION:
p-sub-int-step (*ins*, *p*)

=

p-state (add1-p-pc (*p*),
       p-ctrl-stk (*p*),
       push (tag ('**int**,
               idifference (untag (top1 (p-temp-stk (*p*))),
                             untag (top (p-temp-stk (*p*))))),
            pop (pop (p-temp-stk (*p*)))),
       p-prog-segment (*p*),
       p-data-segment (*p*),
       p-max-ctrl-stk-size (*p*),
       p-max-temp-stk-size (*p*),
       p-word-size (*p*),
       '**run**)

DEFINITION:
p-sub-int-with-carry-okp (*ins*, *p*)

=

    listp (p-temp-stk (*p*))
∧ listp (pop (p-temp-stk (*p*)))
∧ listp (pop (pop (p-temp-stk (*p*))))
∧ p-objectp-type ('**int**, top (p-temp-stk (*p*)), *p*)
∧ p-objectp-type ('**int**, top1 (p-temp-stk (*p*)), *p*)
∧ p-objectp-type ('**bool**, top2 (p-temp-stk (*p*)), *p*)

DEFINITION:
p-sub-int-with-carry-step (*ins*, *p*)
=
p-state (add1-p-pc (*p*),
      p-ctrl-stk (*p*),
      push (tag (**'int**,
                fix-small-integer (
                 idifference (untag (top1 (p-temp-stk (*p*))),
                           iplus (untag (top (p-temp-stk (*p*))),
                                bool-to-nat (
                                  untag (top2 (p-temp-stk (*p*))))))),
                   p-word-size (*p*))),
           push (bool (¬small-integerp (
                     idifference (untag (top1 (p-temp-stk (*p*))),
                             iplus (untag (top (p-temp-stk (*p*))),
                                bool-to-nat (
                                 untag (top2 (p-temp-stk (*p*))))))),
                   p-word-size (*p*))),
              pop (pop (pop (p-temp-stk (*p*))))))),
      p-prog-segment (*p*),
      p-data-segment (*p*),
      p-max-ctrl-stk-size (*p*),
      p-max-temp-stk-size (*p*),
      p-word-size (*p*),
      'run)

DEFINITION:
p-sub-nat-okp (*ins*, *p*)
=
   listp (p-temp-stk (*p*))
∧ listp (pop (p-temp-stk (*p*)))
∧ p-objectp-type (**'nat**, top (p-temp-stk (*p*)), *p*)
∧ p-objectp-type (**'nat**, top1 (p-temp-stk (*p*)), *p*)
∧ (untag (top1 (p-temp-stk (*p*))) ≥ untag (top (p-temp-stk (*p*))))

DEFINITION:
p-sub-nat-step (*ins*, *p*)
=
p-state (add1-p-pc (*p*),
      p-ctrl-stk (*p*),
      push (tag (**'nat**,
             untag (top1 (p-temp-stk (*p*))) − untag (top (p-temp-stk (*p*)))),
          pop (pop (p-temp-stk (*p*)))),
      p-prog-segment (*p*),
      p-data-segment (*p*),
      p-max-ctrl-stk-size (*p*),
      p-max-temp-stk-size (*p*),
      p-word-size (*p*),
      'run)

DEFINITION:
p-sub-nat-with-carry-okp (ins, p)

=

    listp (p-temp-stk (p))
$\wedge$ listp (pop (p-temp-stk (p)))
$\wedge$ listp (pop (pop (p-temp-stk (p))))
$\wedge$ p-objectp-type ($'$**nat**, top (p-temp-stk (p)), p)
$\wedge$ p-objectp-type ($'$**nat**, top1 (p-temp-stk (p)), p)
$\wedge$ p-objectp-type ($'$**bool**, top2 (p-temp-stk (p)), p)

DEFINITION:
p-sub-nat-with-carry-step (ins, p)

=

p-state (add1-p-pc (p),
      p-ctrl-stk (p),
      push (tag ($'$**nat**,
                if   untag (top1 (p-temp-stk (p)))
                  $<$ (untag (top (p-temp-stk (p)))
                     + bool-to-nat (untag (top2 (p-temp-stk (p))))))
              then $2^{\text{p-word-size}\,(p)}$
                  $-$ ((untag (top (p-temp-stk (p)))
                      + bool-to-nat (untag (top2 (p-temp-stk (p))))))
                    $-$ untag (top1 (p-temp-stk (p)))))
              else   untag (top1 (p-temp-stk (p)))
                  $-$ (untag (top (p-temp-stk (p)))
                     + bool-to-nat (untag (top2 (p-temp-stk (p))))))) **endif**),
           push (bool (untag (top1 (p-temp-stk (p)))
                  $<$ (untag (top (p-temp-stk (p)))
                   + bool-to-nat (untag (top2 (p-temp-stk (p)))))),
                pop (pop (pop (p-temp-stk (p)))))))),
      p-prog-segment (p),
      p-data-segment (p),
      p-max-ctrl-stk-size (p),
      p-max-temp-stk-size (p),
      p-word-size (p),
      $'$**run**)

DEFINITION:
p-sub1-int-okp (ins, p)

=

    listp (p-temp-stk (p))
$\wedge$ p-objectp-type ($'$**int**, top (p-temp-stk (p)), p)
$\wedge$ small-integerp (idifference (untag (top (p-temp-stk (p))), **1**), p-word-size (p))

DEFINITION:
p-sub1-int-step (*ins*, *p*)

=

p-state (add1-p-pc (*p*),
   p-ctrl-stk (*p*),
   push (tag (**'int**, idifference (untag (top (p-temp-stk (*p*))), **1**)),
     pop (p-temp-stk (*p*)))),
   p-prog-segment (*p*),
   p-data-segment (*p*),
   p-max-ctrl-stk-size (*p*),
   p-max-temp-stk-size (*p*),
   p-word-size (*p*),
   'run)

DEFINITION:
p-sub1-nat-okp (*ins*, *p*)

=

 listp (p-temp-stk (*p*))
$\wedge$ p-objectp-type (**'nat**, top (p-temp-stk (*p*)), *p*)
$\wedge$ ($\neg$ (untag (top (p-temp-stk (*p*))) $\cong$ **0**))

DEFINITION:
p-sub1-nat-step (*ins*, *p*)

=

p-state (add1-p-pc (*p*),
   p-ctrl-stk (*p*),
   push (tag (**'nat**, untag (top (p-temp-stk (*p*))) −1), pop (p-temp-stk (*p*)))),
   p-prog-segment (*p*),
   p-data-segment (*p*),
   p-max-ctrl-stk-size (*p*),
   p-max-temp-stk-size (*p*),
   p-word-size (*p*),
   'run)

DEFINITION:
p-test-and-jump-okp (*ins*, *type*, *test*, *p*)

=

listp (p-temp-stk (*p*)) $\wedge$ p-objectp-type (*type*, top (p-temp-stk (*p*)), *p*)

DEFINITION:
p-test-and-jump-step (*test*, *lab*, *p*)

=

if *test*
 then p-state (pc (*lab*, p-current-program (*p*)),
                   p-ctrl-stk (*p*),
                   pop (p-temp-stk (*p*)),
                   p-prog-segment (*p*),
                   p-data-segment (*p*),
                   p-max-ctrl-stk-size (*p*),
                   p-max-temp-stk-size (*p*),
                   p-word-size (*p*),
                   'run)
 else p-state (add1-p-pc (*p*),
                   p-ctrl-stk (*p*),
                   pop (p-temp-stk (*p*)),
                   p-prog-segment (*p*),
                   p-data-segment (*p*),
                   p-max-ctrl-stk-size (*p*),
                   p-max-temp-stk-size (*p*),
                   p-word-size (*p*),
                   'run) **endif**

DEFINITION:
p-test-bitv-and-jump-okp (*ins*, *p*)

=

p-test-and-jump-okp (*ins*,
                     'bitv,
                     p-test-bitvp (cadr (*ins*), untag (top (p-temp-stk (*p*)))),
                     p)

DEFINITION:
p-test-bitv-and-jump-step (*ins*, *p*)

=

p-test-and-jump-step (p-test-bitvp (cadr (*ins*), untag (top (p-temp-stk (*p*)))),
                      caddr (*ins*),
                      p)

DEFINITION:
p-test-bitvp (*flg*, *x*)

=

if *flg* = **'all-zero** **then** all-zero-bitvp (*x*)
 else ¬ all-zero-bitvp (*x*) **endif**

DEFINITION:
p-test-bool-and-jump-okp *(ins, p)*

=

p-test-and-jump-okp *(ins,*
     'bool,
     p-test-boolp (cadr *(ins)*, untag (top (p-temp-stk *(p)*)))),
     p)

DEFINITION:
p-test-bool-and-jump-step *(ins, p)*

=

p-test-and-jump-step (p-test-boolp (cadr *(ins)*, untag (top (p-temp-stk *(p)*)))),
     caddr *(ins)*,
     p)

DEFINITION:
p-test-boolp *(flg, x)*

=

if *flg* = **'t** **then** $x =$ **'t**
 else $x =$ **'f** **endif**

DEFINITION:
p-test-int-and-jump-okp *(ins, p)*

=

p-test-and-jump-okp *(ins,*
     'int,
     p-test-intp (cadr *(ins)*, untag (top (p-temp-stk *(p)*)))),
     p)

DEFINITION:
p-test-int-and-jump-step *(ins, p)*

=

p-test-and-jump-step (p-test-intp (cadr *(ins)*, untag (top (p-temp-stk *(p)*)))),
     caddr *(ins)*,
     p)

DEFINITION:
p-test-intp *(flg, x)*

=

case on *flg*:
 case = **zero** **then** $x = 0$
 case = **not-zero** **then** $x \neq 0$
 case = **neg** **then** negativep (x)
 case = **not-neg** **then** $\neg$ negativep (x)
 case = **pos** **then** $(x \in N) \wedge (x \neq 0)$
 otherwise $(x = 0) \vee$ negativep (x) **endcase**

DEFINITION:
p-test-nat-and-jump-okp (*ins*, *p*)

=

p-test-and-jump-okp (*ins*,
                      'nat,
                      p-test-natp (cadr (*ins*), untag (top (p-temp-stk (*p*)))),
                      p)

DEFINITION:
p-test-nat-and-jump-step (*ins*, *p*)

=

p-test-and-jump-step (p-test-natp (cadr (*ins*), untag (top (p-temp-stk (*p*)))),
                       caddr (*ins*),
                       p)

DEFINITION:
p-test-natp (*flg*, *x*)

=

if *flg* = **'zero** **then** $x = 0$
else $x \neq 0$ **endif**

DEFINITION:
p-xor-bitv-okp (*ins*, *p*)

=

    listp (p-temp-stk (*p*))
∧ listp (pop (p-temp-stk (*p*)))
∧ p-objectp-type (**'bitv**, top (p-temp-stk (*p*)), *p*)
∧ p-objectp-type (**'bitv**, top1 (p-temp-stk (*p*)), *p*)

DEFINITION:
p-xor-bitv-step (*ins*, *p*)

=

p-state (add1-p-pc (*p*),
        p-ctrl-stk (*p*),
        push (tag (**'bitv**,
                   xor-bitv (untag (top1 (p-temp-stk (*p*))),
                             untag (top (p-temp-stk (*p*)))))),
              pop (pop (p-temp-stk (*p*)))),
        p-prog-segment (*p*),
        p-data-segment (*p*),
        p-max-ctrl-stk-size (*p*),
        p-max-temp-stk-size (*p*),
        p-word-size (*p*),
        'run)

DEFINITION:
pair-formal-vars-with-actuals (*formal-vars*, *temp-stk*)
=
pairlist (*formal-vars*, rev (first-n (length (*formal-vars*), *temp-stk*)))

DEFINITION:
pair-temps-with-initial-values (*temp-var-dcls*)
=
if nlistp (*temp-var-dcls*) **then nil**
 else cons (cons (caar (*temp-var-dcls*), cadar (*temp-var-dcls*)),
            pair-temps-with-initial-values (cdr (*temp-var-dcls*))) **endif**

DEFINITION:
pc (*lab*, *program*)
=
tag ('**pc**, cons (name (*program*), find-label (*lab*, program-body (*program*))))

DEFINITION:
pcpp (*x*, *segment*)
=
   listp (*x*)
$\wedge$ (adp-offset (*x*) $\in$ **N**)
$\wedge$ definedp (adp-name (*x*), *segment*)
$\wedge$ (adp-offset (*x*) < length (program-body (definition (adp-name (*x*), *segment*))))

DEFINITION:
pop (*stk*) = cdr (*stk*)

DEFINITION:
popn (*n*, *x*)
=
if $n \cong 0$ **then** *x*
 else popn (*n* −1, cdr (*x*)) **endif**

DEFINITION:
program-body (*d*) = cdddr (*d*)

DEFINITION:
proper-labeled-p-instructionsp (*lst*, *name*, *p*)
=
if nlistp (*lst*) **then** *lst* = **nil**
 else   legal-labelp (car (*lst*))
      $\wedge$ proper-p-instructionp (unlabel (car (*lst*)), *name*, *p*)
      $\wedge$ proper-labeled-p-instructionsp (cdr (*lst*), *name*, *p*) **endif**

DEFINITION:
proper-p-add-addr-instructionp (*ins*, *name*, *p*) = length (*ins*) = **1**

DEFINITION:
proper-p-add-int-instructionp (*ins*, *name*, *p*) = length (*ins*) = **1**

DEFINITION:
proper-p-add-int-with-carry-instructionp (*ins*, *name*, *p*)
=
length (*ins*) = **1**

DEFINITION:
proper-p-add-nat-instructionp (*ins*, *name*, *p*) = length (*ins*) = **1**

DEFINITION:
proper-p-add-nat-with-carry-instructionp (*ins*, *name*, *p*)
=
length (*ins*) = **1**

DEFINITION:
proper-p-add1-int-instructionp (*ins*, *name*, *p*) = length (*ins*) = **1**

DEFINITION:
proper-p-add1-nat-instructionp (*ins*, *name*, *p*) = length (*ins*) = **1**

DEFINITION:
proper-p-alistp (*alist*, *p*)
=
if nlistp (*alist*) **then** *alist* = **nil**
  else   listp (car (*alist*))
        ∧   litatom (caar (*alist*))
        ∧   p-objectp (cdar (*alist*), *p*)
        ∧   proper-p-alistp (cdr (*alist*), *p*) **endif**

DEFINITION:
proper-p-and-bitv-instructionp (*ins*, *name*, *p*) = length (*ins*) = **1**

DEFINITION:
proper-p-and-bool-instructionp (*ins*, *name*, *p*) = length (*ins*) = **1**

DEFINITION:
proper-p-area (*area*, *p*)
=
litatom (car (*area*)) ∧ listp (cdr (*area*)) ∧ all-p-objectps (cdr (*area*), *p*)

DEFINITION:
proper-p-call-instructionp (*ins*, *name*, *p*)
=
(length (*ins*) = **2**) $\wedge$ definedp (cadr (*ins*), p-prog-segment (*p*))

DEFINITION:
proper-p-ctrl-stkp (*ctrl-stk*, *name*, *p*)
=
if nlistp (*ctrl-stk*) **then** *ctrl-stk* = **nil**
else   proper-p-framep (top (*ctrl-stk*), *name*, *p*)
   $\wedge$ proper-p-ctrl-stkp (pop (*ctrl-stk*),
                             area-name (ret-pc (top (*ctrl-stk*))),
                             p) **endif**

DEFINITION:
proper-p-data-segmentp (*data-segment*, *p*)
=
if nlistp (*data-segment*) **then** *data-segment* = **nil**
else   proper-p-area (car (*data-segment*), *p*)
   $\wedge$ ($\neg$ definedp (caar (*data-segment*), cdr (*data-segment*)))
   $\wedge$ proper-p-data-segmentp (cdr (*data-segment*), *p*) **endif**

DEFINITION:
proper-p-deposit-instructionp (*ins*, *name*, *p*) = length (*ins*) = **1**

DEFINITION:
proper-p-deposit-temp-stk-instructionp (*ins*, *name*, *p*)
=
length (*ins*) = **1**

DEFINITION:
proper-p-div2-nat-instructionp (*ins*, *name*, *p*) = length (*ins*) = **1**

DEFINITION:
proper-p-eq-instructionp (*ins*, *name*, *p*) = length (*ins*) = **1**

DEFINITION:
proper-p-fetch-instructionp (*ins*, *name*, *p*) = length (*ins*) = **1**

DEFINITION:
proper-p-fetch-temp-stk-instructionp (*ins*, *name*, *p*)
=
length (*ins*) = **1**

DEFINITION:
proper-p-framep (*frame*, *name*, *p*)
=

    listp (*frame*)
∧ listp (cdr (*frame*))
∧ (cddr (*frame*) = **nil**)
∧ proper-p-alistp (bindings (*frame*), *p*)
∧ (strip-cars (bindings (*frame*))
    = local-vars (definition (*name*, p-prog-segment (*p*))))
∧ p-objectp-type (**'pc**, ret-pc (*frame*), *p*)

DEFINITION:
proper-p-instructionp (*ins*, *name*, *p*)
=

    properp (*ins*)
∧ **case on** car (*ins*):
    case = **call**
    then proper-p-call-instructionp (*ins*, *name*, *p*)
    case = **ret**
    then proper-p-ret-instructionp (*ins*, *name*, *p*)
    case = **locn**
    then proper-p-locn-instructionp (*ins*, *name*, *p*)
    case = **push-constant**
    then proper-p-push-constant-instructionp (*ins*, *name*, *p*)
    case = **push-local**
    then proper-p-push-local-instructionp (*ins*, *name*, *p*)
    case = **push-global**
    then proper-p-push-global-instructionp (*ins*, *name*, *p*)
    case = **push-ctrl-stk-free-size**
    then proper-p-push-ctrl-stk-free-size-instructionp (*ins*, *name*, *p*)
    case = **push-temp-stk-free-size**
    then proper-p-push-temp-stk-free-size-instructionp (*ins*, *name*, *p*)
    case = **push-temp-stk-index**
    then proper-p-push-temp-stk-index-instructionp (*ins*, *name*, *p*)
    case = **jump-if-temp-stk-full**
    then proper-p-jump-if-temp-stk-full-instructionp (*ins*, *name*, *p*)
    case = **jump-if-temp-stk-empty**
    then proper-p-jump-if-temp-stk-empty-instructionp (*ins*, *name*, *p*)
    case = **pop**
    then proper-p-pop-instructionp (*ins*, *name*, *p*)
    case = **pop***
    then proper-p-pop*-instructionp (*ins*, *name*, *p*)
    case = **popn**
    then proper-p-popn-instructionp (*ins*, *name*, *p*)
    case = **pop-local**
    then proper-p-pop-local-instructionp (*ins*, *name*, *p*)
    case = **pop-global**
    then proper-p-pop-global-instructionp (*ins*, *name*, *p*)
    case = **pop-locn**

then proper-p-pop-locn-instructionp (*ins, name, p*)

case = **pop-call**

then proper-p-pop-call-instructionp (*ins, name, p*)

case = **fetch-temp-stk**

then proper-p-fetch-temp-stk-instructionp (*ins, name, p*)

case = **deposit-temp-stk**

then proper-p-deposit-temp-stk-instructionp (*ins, name, p*)

case = **jump**

then proper-p-jump-instructionp (*ins, name, p*)

case = **jump-case**

then proper-p-jump-case-instructionp (*ins, name, p*)

case = **pushj**

then proper-p-pushj-instructionp (*ins, name, p*)

case = **popj**

then proper-p-popj-instructionp (*ins, name, p*)

case = **set-local**

then proper-p-set-local-instructionp (*ins, name, p*)

case = **set-global**

then proper-p-set-global-instructionp (*ins, name, p*)

case = **test-nat-and-jump**

then proper-p-test-nat-and-jump-instructionp (*ins, name, p*)

case = **test-int-and-jump**

then proper-p-test-int-and-jump-instructionp (*ins, name, p*)

case = **test-bool-and-jump**

then proper-p-test-bool-and-jump-instructionp (*ins, name, p*)

case = **test-bitv-and-jump**

then proper-p-test-bitv-and-jump-instructionp (*ins, name, p*)

case = **no-op**

then proper-p-no-op-instructionp (*ins, name, p*)

case = **add-addr**

then proper-p-add-addr-instructionp (*ins, name, p*)

case = **sub-addr**

then proper-p-sub-addr-instructionp (*ins, name, p*)

case = **eq**

then proper-p-eq-instructionp (*ins, name, p*)

case = **lt-addr**

then proper-p-lt-addr-instructionp (*ins, name, p*)

case = **fetch**

then proper-p-fetch-instructionp (*ins, name, p*)

case = **deposit**

then proper-p-deposit-instructionp (*ins, name, p*)

case = **add-int**

then proper-p-add-int-instructionp (*ins, name, p*)

case = **add-int-with-carry**

then proper-p-add-int-with-carry-instructionp (*ins, name, p*)

case = **add1-int**

then proper-p-add1-int-instructionp (*ins, name, p*)

case = **sub-int**

then proper-p-sub-int-instructionp (*ins, name, p*)

case = **sub-int-with-carry**

then proper-p-sub-int-with-carry-instructionp (*ins, name, p*)
case = sub1-int
then proper-p-sub1-int-instructionp (*ins, name, p*)
case = neg-int
then proper-p-neg-int-instructionp (*ins, name, p*)
case = lt-int
then proper-p-lt-int-instructionp (*ins, name, p*)
case = int-to-nat
then proper-p-int-to-nat-instructionp (*ins, name, p*)
case = add-nat
then proper-p-add-nat-instructionp (*ins, name, p*)
case = add-nat-with-carry
then proper-p-add-nat-with-carry-instructionp (*ins, name, p*)
case = add1-nat
then proper-p-add1-nat-instructionp (*ins, name, p*)
case = sub-nat
then proper-p-sub-nat-instructionp (*ins, name, p*)
case = sub-nat-with-carry
then proper-p-sub-nat-with-carry-instructionp (*ins, name, p*)
case = sub1-nat
then proper-p-sub1-nat-instructionp (*ins, name, p*)
case = lt-nat
then proper-p-lt-nat-instructionp (*ins, name, p*)
case = mult2-nat
then proper-p-mult2-nat-instructionp (*ins, name, p*)
case = mult2-nat-with-carry-out
then proper-p-mult2-nat-with-carry-out-instructionp (*ins, name, p*)
case = div2-nat
then proper-p-div2-nat-instructionp (*ins, name, p*)
case = or-bitv
then proper-p-or-bitv-instructionp (*ins, name, p*)
case = and-bitv
then proper-p-and-bitv-instructionp (*ins, name, p*)
case = not-bitv
then proper-p-not-bitv-instructionp (*ins, name, p*)
case = xor-bitv
then proper-p-xor-bitv-instructionp (*ins, name, p*)
case = rsh-bitv
then proper-p-rsh-bitv-instructionp (*ins, name, p*)
case = lsh-bitv
then proper-p-lsh-bitv-instructionp (*ins, name, p*)
case = or-bool
then proper-p-or-bool-instructionp (*ins, name, p*)
case = and-bool
then proper-p-and-bool-instructionp (*ins, name, p*)
case = not-bool
then proper-p-not-bool-instructionp (*ins, name, p*)
otherwise f endcase

DEFINITION:
proper-p-int-to-nat-instructionp (*ins*, *name*, *p*) = length (*ins*) = **1**

DEFINITION:
proper-p-jump-case-instructionp (*ins*, *name*, *p*)
=
    listp (cdr (*ins*))
∧ all-find-labelp (cdr (*ins*),
                        program-body (definition (*name*, p-prog-segment (*p*))))

DEFINITION:
proper-p-jump-if-temp-stk-empty-instructionp (*ins*, *name*, *p*)
=
    (length (*ins*) = **2**)
∧ find-labelp (cadr (*ins*), program-body (definition (*name*, p-prog-segment (*p*))))

DEFINITION:
proper-p-jump-if-temp-stk-full-instructionp (*ins*, *name*, *p*)
=
    (length (*ins*) = **2**)
∧ find-labelp (cadr (*ins*), program-body (definition (*name*, p-prog-segment (*p*))))

DEFINITION:
proper-p-jump-instructionp (*ins*, *name*, *p*)
=
    (length (*ins*) = **2**)
∧ find-labelp (cadr (*ins*), program-body (definition (*name*, p-prog-segment (*p*))))

DEFINITION:
proper-p-locn-instructionp (*ins*, *name*, *p*)
=
    (length (*ins*) = **2**)
∧ (cadr (*ins*) ∈ local-vars (definition (*name*, p-prog-segment (*p*))))

DEFINITION:
proper-p-lsh-bitv-instructionp (*ins*, *name*, *p*) = length (*ins*) = **1**

DEFINITION:
proper-p-lt-addr-instructionp (*ins*, *name*, *p*) = length (*ins*) = **1**

DEFINITION:
proper-p-lt-int-instructionp (*ins*, *name*, *p*) = length (*ins*) = **1**

DEFINITION:
proper-p-lt-nat-instructionp (*ins*, *name*, *p*) = length (*ins*) = **1**

DEFINITION:
proper-p-mult2-nat-instructionp (*ins*, *name*, *p*) = length (*ins*) = **1**

DEFINITION:
proper-p-mult2-nat-with-carry-out-instructionp (*ins*, *name*, *p*)
=
length (*ins*) = **1**

DEFINITION:
proper-p-neg-int-instructionp (*ins*, *name*, *p*) = length (*ins*) = **1**

DEFINITION:
proper-p-no-op-instructionp (*ins*, *name*, *p*) = length (*ins*) = **1**

DEFINITION:
proper-p-not-bitv-instructionp (*ins*, *name*, *p*) = length (*ins*) = **1**

DEFINITION:
proper-p-not-bool-instructionp (*ins*, *name*, *p*) = length (*ins*) = **1**

DEFINITION:
proper-p-or-bitv-instructionp (*ins*, *name*, *p*) = length (*ins*) = **1**

DEFINITION:
proper-p-or-bool-instructionp (*ins*, *name*, *p*) = length (*ins*) = **1**

DEFINITION:
proper-p-pop*-instructionp (*ins*, *name*, *p*)
=
(length (*ins*) = **2**) ∧ small-naturalp (cadr (*ins*), p-word-size (*p*))

DEFINITION:
proper-p-pop-call-instructionp (*ins*, *name*, *p*) = length (*ins*) = **1**

DEFINITION:
proper-p-pop-global-instructionp (*ins*, *name*, *p*)
=
(length (*ins*) = **2**) ∧ definedp (cadr (*ins*), p-data-segment (*p*))

DEFINITION:
proper-p-pop-instructionp (*ins*, *name*, *p*) = length (*ins*) = **1**

DEFINITION:
proper-p-pop-local-instructionp (*ins*, *name*, *p*)
$$=$$
$$(\text{length}\,(\textit{ins}) = \mathbf{2})$$
$$\wedge\ (\text{cadr}\,(\textit{ins}) \in \text{local-vars}\,(\text{definition}\,(\textit{name},\ \text{p-prog-segment}\,(p))))$$

DEFINITION:
proper-p-pop-locn-instructionp (*ins*, *name*, *p*)
$$=$$
$$(\text{length}\,(\textit{ins}) = \mathbf{2})$$
$$\wedge\ (\text{cadr}\,(\textit{ins}) \in \text{local-vars}\,(\text{definition}\,(\textit{name},\ \text{p-prog-segment}\,(p))))$$

DEFINITION:
proper-p-popj-instructionp (*ins*, *name*, *p*) = length (*ins*) = $\mathbf{1}$

DEFINITION:
proper-p-popn-instructionp (*ins*, *name*, *p*) = length (*ins*) = $\mathbf{1}$

DEFINITION:
proper-p-prog-segmentp (*segment*, *p*)
$$=$$
if nlistp (*segment*) **then** *segment* = **nil**
else   proper-p-programp (car (*segment*), *p*)
    $\wedge$ proper-p-prog-segmentp (cdr (*segment*), *p*) **endif**

DEFINITION:
proper-p-program-bodyp (*lst*, *name*, *p*)
$$=$$
listp (*lst*)
$\wedge$ proper-labeled-p-instructionsp (*lst*, *name*, *p*)
$\wedge$ fall-off-proofp (*lst*)

DEFINITION:
proper-p-programp (*prog*, *p*)
$$=$$
litatom (name (*prog*))
$\wedge$ all-litatoms (formal-vars (*prog*))
$\wedge$ proper-p-temp-var-dclsp (temp-var-dcls (*prog*), *p*)
$\wedge$ proper-p-program-bodyp (program-body (*prog*), name (*prog*), *p*)

DEFINITION:
proper-p-push-constant-instructionp (*ins*, *name*, *p*)
=

 (length (*ins*) = **2**)
∧ (p-objectp (cadr (*ins*), *p*)
  ∨ (cadr (*ins*) = **'pc**)
  ∨ find-labelp (cadr (*ins*),
        program-body (definition (*name*, p-prog-segment (*p*))))))

DEFINITION:
proper-p-push-ctrl-stk-free-size-instructionp (*ins*, *name*, *p*)
=

length (*ins*) = **1**

DEFINITION:
proper-p-push-global-instructionp (*ins*, *name*, *p*)
=

(length (*ins*) = **2**) ∧ definedp (cadr (*ins*), p-data-segment (*p*))

DEFINITION:
proper-p-push-local-instructionp (*ins*, *name*, *p*)
=

 (length (*ins*) = **2**)
∧ (cadr (*ins*) ∈ local-vars (definition (*name*, p-prog-segment (*p*))))

DEFINITION:
proper-p-push-temp-stk-free-size-instructionp (*ins*, *name*, *p*)
=

length (*ins*) = **1**

DEFINITION:
proper-p-push-temp-stk-index-instructionp (*ins*, *name*, *p*)
=

(length (*ins*) = **2**) ∧ small-naturalp (cadr (*ins*), p-word-size (*p*))

DEFINITION:
proper-p-pushj-instructionp (*ins*, *name*, *p*)
=

 (length (*ins*) = **2**)
∧ find-labelp (cadr (*ins*), program-body (definition (*name*, p-prog-segment (*p*))))

DEFINITION:
proper-p-ret-instructionp (*ins*, *name*, *p*) = length (*ins*) = **1**

DEFINITION:
proper-p-rsh-bitv-instructionp (*ins*, *name*, *p*) = length (*ins*) = **1**

DEFINITION:
proper-p-set-global-instructionp (*ins*, *name*, *p*)

=

(length (*ins*) = **2**) ∧ definedp (cadr (*ins*), p-data-segment (*p*))

DEFINITION:
proper-p-set-local-instructionp (*ins*, *name*, *p*)

=

    (length (*ins*) = **2**)
∧ (cadr (*ins*) ∈ local-vars (definition (*name*, p-prog-segment (*p*))))

DEFINITION:
proper-p-statep (*p*)

=

    p-statep (*p*)
∧ p-objectp-type (**'pc**, p-pc (*p*), *p*)
∧ listp (p-ctrl-stk (*p*))
∧ proper-p-framep (top (p-ctrl-stk (*p*)), area-name (p-pc (*p*)), *p*)
∧ proper-p-ctrl-stkp (pop (p-ctrl-stk (*p*)),
                              area-name (ret-pc (top (p-ctrl-stk (*p*)))),
                              p)
∧ (p-max-ctrl-stk-size (*p*) ≥ p-ctrl-stk-size (p-ctrl-stk (*p*)))
∧ proper-p-temp-stkp (p-temp-stk (*p*), *p*)
∧ (p-max-temp-stk-size (*p*) ≥ length (p-temp-stk (*p*)))
∧ proper-p-prog-segmentp (p-prog-segment (*p*), *p*)
∧ proper-p-data-segmentp (p-data-segment (*p*), *p*)
∧ (p-max-ctrl-stk-size (*p*) ∈ **N**)
∧ (p-max-temp-stk-size (*p*) ∈ **N**)
∧ (p-word-size (*p*) ∈ **N**)
∧ (p-max-ctrl-stk-size (*p*) < $2^{\text{p-word-size }(p)}$)
∧ (p-max-temp-stk-size (*p*) < $2^{\text{p-word-size }(p)}$)
∧ (**0** < p-word-size (*p*))

DEFINITION:
proper-p-sub-addr-instructionp (*ins*, *name*, *p*) = length (*ins*) = **1**

DEFINITION:
proper-p-sub-int-instructionp (*ins*, *name*, *p*) = length (*ins*) = **1**

DEFINITION:
proper-p-sub-int-with-carry-instructionp (*ins*, *name*, *p*)

=

length (*ins*) = **1**

DEFINITION:
proper-p-sub-nat-instructionp (*ins*, *name*, *p*) = length (*ins*) = **1**

DEFINITION:
proper-p-sub-nat-with-carry-instructionp (*ins, name, p*)
=
length (*ins*) = **1**

DEFINITION:
proper-p-sub1-int-instructionp (*ins, name, p*) = length (*ins*) = **1**

DEFINITION:
proper-p-sub1-nat-instructionp (*ins, name, p*) = length (*ins*) = **1**

DEFINITION:
proper-p-temp-stkp (*temp-stk, p*)
=
if nlistp (*temp-stk*) **then** *temp-stk* = **nil**
  else p-objectp (top (*temp-stk*), p) ∧ proper-p-temp-stkp (pop (*temp-stk*), p) **endif**

DEFINITION:
proper-p-temp-var-dclsp (*temp-var-dcls, p*)
=
if nlistp (*temp-var-dcls*) **then t**
  else   litatom (caar (*temp-var-dcls*))
      ∧ p-objectp (cadar (*temp-var-dcls*), p)
      ∧ proper-p-temp-var-dclsp (cdr (*temp-var-dcls*), p) **endif**

DEFINITION:
proper-p-test-bitv-and-jump-instructionp (*ins, name, p*)
=
    (length (*ins*) = **3**)
∧ find-labelp (caddr (*ins*),
                program-body (definition (*name*, p-prog-segment (*p*))))

DEFINITION:
proper-p-test-bool-and-jump-instructionp (*ins, name, p*)
=
    (length (*ins*) = **3**)
∧ find-labelp (caddr (*ins*),
                program-body (definition (*name*, p-prog-segment (*p*))))

DEFINITION:
proper-p-test-int-and-jump-instructionp (*ins, name, p*)
=
    (length (*ins*) = **3**)
∧ find-labelp (caddr (*ins*),
                program-body (definition (*name*, p-prog-segment (*p*))))

DEFINITION:
proper-p-test-nat-and-jump-instructionp (*ins*, *name*, *p*)
=

 (length (*ins*) = **3**)
∧ find-labelp (caddr (*ins*),
       program-body (definition (*name*, p-prog-segment (*p*)))))

DEFINITION:
proper-p-xor-bitv-instructionp (*ins*, *name*, *p*) = length (*ins*) = **1**

DEFINITION:
properp (*l*)
=

if listp (*l*) **then** properp (cdr (*l*))
 else *l* = **nil** **endif**

DEFINITION:
push (*x*, *stk*) = cons (*x*, *stk*)

DEFINITION:
put (*val*, *n*, *lst*)
=

if *n* ≅ **0**
 then if listp (*lst*) **then** cons (*val*, cdr (*lst*))
   else list (*val*) **endif**
 else cons (car (*lst*), put (*val*,*n* −1, cdr (*lst*))) **endif**

DEFINITION:
put-assoc (*val*, *name*, *alist*)
=

if nlistp (*alist*) **then** *alist*
 elseif *name* = caar (*alist*) **then** cons (cons (*name*, *val*), cdr (*alist*))
 else cons (car (*alist*), put-assoc (*val*, *name*, cdr (*alist*))) **endif**

DEFINITION:
put-value (*val*, *name*, *alist*) = put-assoc (*val*, *name*, *alist*)

DEFINITION:
put-value-indirect (*val*, *n*, *lst*)
=

if listp (*lst*)
 then if *n* ≅ **0** **then** cons (cons (caar (*lst*), *val*), cdr (*lst*))
   else cons (car (*lst*), put-value-indirect (*val*,*n* −1, cdr (*lst*))) **endif**
 else *lst* **endif**

DEFINITION:
ret-pc (*frame*) = cadr (*frame*)

DEFINITION:
rev (*x*)

=

if nlistp (*x*) **then nil**
 else append (rev (cdr (*x*)), list (car (*x*))) **endif**

DEFINITION:
rget (*n*, *lst*) = get ((length (*lst*) − *n*) −1, *lst*)

DEFINITION:
rput (*val*, *n*, *lst*) = put (*val*,(length (*lst*) − *n*) −1, *lst*)

DEFINITION:
rsh-bitv (*a*) = cons (**0**, all-but-last (*a*))

DEFINITION:
set-local-var-indirect (*val*, *index*, *ctrl-stk*)

=

push (p-frame (put-value-indirect (*val*, *index*, bindings (top (*ctrl-stk*))),
              ret-pc (top (*ctrl-stk*))),
     pop (*ctrl-stk*))

DEFINITION:
set-local-var-value (*val*, *var*, *ctrl-stk*)

=

push (p-frame (put-value (*val*, *var*, bindings (top (*ctrl-stk*))),
              ret-pc (top (*ctrl-stk*))),
     pop (*ctrl-stk*))

DEFINITION:
small-integerp (*i*, *word-size*)

=

    integerp (*i*)
$\wedge$ ($\neg$ ilessp (*i*, $- 2^{(word\text{-}size\,-1)}$))
$\wedge$ ilessp (*i*, $2^{word\text{-}size\,-1}$)

DEFINITION:
small-naturalp (*i*, *word-size*) = ($i \in \mathbf{N}$) $\wedge$ ($i < 2^{word\text{-}size}$)

DEFINITION:
sub-addr (*addr*, *n*) = tag (type (*addr*), sub-adp (untag (*addr*), *n*))

DEFINITION:
sub-adp (adp, n) = cons (adp-name (adp), adp-offset $(adp) - n$)

DEFINITION:
tag $(type, obj)$ = list $(type, obj)$

DEFINITION:
temp-var-dcls (d) = caddr (d)

DEFINITION:
top (stk) = car (stk)

DEFINITION:
top1 (stk) = top (pop (stk))

DEFINITION:
top2 (stk) = top (pop (pop (stk)))

DEFINITION:
type $(const)$ = car $(const)$

DEFINITION:
unabbreviate-constant (c, p)

=

if c = **'pc** **then** add1-p-pc (p)
 elseif nlistp (c) **then** pc $(c$, p-current-program $(p))$
 else c **endif**

DEFINITION:
unlabel (x)

=

if labeledp (x) **then** cadddr (x)
 else x **endif**

DEFINITION:
untag $(const)$ = cadr $(const)$

DEFINITION:
x-y-error-msg (x, y)

=

pack (append (unpack (**'illegal-**),
                append (unpack (y), cdr (unpack (**'g-instruction**)))))

DEFINITION:
xor-bit (*bit1*, *bit2*)
=

if *bit1* = 0
 then if *bit2* = 0 **then** 0
      else 1 **endif**
 elseif *bit2* = 0 **then** 1
 else 0 **endif**

DEFINITION:
xor-bitv (*a*, *b*)
=

if nlistp (*a*) **then nil**
 else cons (xor-bit (car (*a*), car (*b*)), xor-bitv (cdr (*a*), cdr (*b*))) **endif**

Appendix III

The Formal Definition of FM9001

In this appendix we present the formal definition of the FM9001. As before, the definitions are presented in alphabetical order and are fully indexed and cross-indexed.

III.1. A Guide to the Formal Definition of FM9001

The state of the FM9001 is formally represented by a list structure of the form
'((*regs* (*z n v c*)) *mem*) where

- *regs* is a ram tree of depth **4** specifying the values of the 16 registers, all of which are marked 'ram'.

- *z, n, v,* and *c* are the Boolean values of the respective flags.

- *mem* is a stubbed out ram tree of depth **32**.

The formal definition of FM9001 is

DEFINITION:
fm9001 (*state, n*)
=
if $n \cong 0$ **then** *state*
 else fm9001 (fm9001-step (*state*, nat-to-v (**15**, reg-size)),*n* −1) **endif**

where *state* is an FM9001 state and *n* is a natural number indicating how many machine instructions to execute. Observe that 'fm9001' simply applies the single stepper, 'fm9001-step', *n* times.

The single stepper is just one iteration through the fetch-execute cycle described informally on page 73. The single stepper takes two arguments, the *state* and the bit vector representing the number of the register to use as the program counter,

DEFINITION:
fm9001-step (*state, pc-reg*)
=
fm9001-fetch (regs (car (*state*)), flags (car (*state*)), cadr (*state*), *pc-reg*)

and calls 'fm9001-fetch' after decomposing *state* into its registers, flags, and memory components.

The first step in the fetch-execute cycle is to fetch the current instruction and increment the program counter. Here is the formal definition.

DEFINITION:
fm9001-fetch (*regs*, *flags*, *mem*, *pc-reg*)
=
fm9001-operand-a (write-mem (*pc-reg*, *regs*, v-inc (read-mem (*pc-reg*, *regs*))),
     flags,
     mem,
     read-mem (read-mem (*pc-reg*, *regs*), *mem*))

The expression read-mem (*pc-reg*, *regs*) fetches the contents of the *pc* register from the ram tree representing the register file, obtaining the address of the current instruction. Then the immediately superior expression, read-mem (read-mem (*pc-reg*, *regs*), *mem*), fetches the current instruction from the ram tree representing memory. The 'write-mem' expression above increments the *pc* register by one. Finally, the incremented register file, the *flags*, *mem*, and the current instruction are passed to 'fm9001-operand-a', the next step in the fetch execute cycle.

Recall that the next step is to obtain operand A and update the registers as necessary given the addressing mode for operand A in the current instruction, *ins*. Here is the formal definition.

DEFINITION:
fm9001-operand-a (*regs*, *flags*, *mem*, *ins*)
=
fm9001-operand-b (**if** a-immediate-p (*ins*) **then** *regs*
     elseif pre-dec-p (mode-a (*ins*))
     then write-mem (rn-a (*ins*),
          regs,
          v-dec (read-mem (rn-a (*ins*), *regs*)))
     elseif post-inc-p (mode-a (*ins*))
     then write-mem (rn-a (*ins*),
          regs,
          v-inc (read-mem (rn-a (*ins*), *regs*)))
     else *regs* **endif**,
    flags,
    mem,
    ins,
    if a-immediate-p (*ins*)
    then sign-extend (a-immediate (*ins*), **32**)
    elseif reg-direct-p (mode-a (*ins*))
    then read-mem (rn-a (*ins*), *regs*)
    elseif pre-dec-p (mode-a (*ins*))
    then read-mem (v-dec (read-mem (rn-a (*ins*), *regs*)), *mem*)
    else read-mem (read-mem (rn-a (*ins*), *regs*), *mem*) **endif**)

Observe that this function calls 'fm9001-operand-b' after updating the register file as required by the addressing mode for operand A. It passes 'fm9001-operand-b', the last argument, the value obtained for operand A.

The definition of the fetch-execute cycle continues through 'fm9001-operand-b' to 'fm9001-alu-operation', where the next state is constructed. We leave it to the reader to pursue the formal definitions using the Index and the alphabetized listing that follows. Readers uninterested in pursuing the formal definition of the FM9001 at this time should skip to page 259.

III.2. Alphabetical Listing of the FM9001 Definitions

DEFINITION:
a-immediate (*instruction*) = subrange (*instruction*, **0**, **8**)

DEFINITION:
a-immediate-p (*instruction*) = nth (**9**, *instruction*)

DEFINITION:
b-and (*a*, *b*) = $a \wedge b$

DEFINITION:
b-buf (*x*)

=

if *x* then t
else f endif

DEFINITION:
b-equv (*x*, *y*)

=

if *x*
then if *y* then t
          else f endif
elseif *y* then f
else t endif

DEFINITION:
b-if (*c*, *a*, *b*)

=

if *c*
then if *a* then t
          else f endif
elseif *b* then t
else f endif

DEFINITION:
b-not (*x*) = $\neg x$

DEFINITION:
b-or $(a, b) = a \vee b$

DEFINITION:
b-xor (x, y)
=
if x
  then if y **then f**
        else t endif
  elseif y **then t**
  else f endif

DEFINITION:
b-xor3 $(a, b, c) = $ b-xor (b-xor $(a, b), c)$

DEFINITION:
boolfix (x)
=
if x **then t**
else f endif

DEFINITION:
bv $(cvzbv) = $ cdddr $(cvzbv)$

DEFINITION:
c $(cvzbv) = $ car $(cvzbv)$

DEFINITION:
c-flag $(flags) = $ nth $(\mathbf{3}, flags)$

DEFINITION:
c-set $(set\text{-}flags) = $ nth $(\mathbf{3}, set\text{-}flags)$

DEFINITION:
cvzbv $(carry, overflow, vector)$
=
cons $(carry,$ cons $(overflow,$ cons (v-zerop $(vector), vector)))$

DEFINITION:
cvzbv-dec $(a) = $ cvzbv-v-subtracter $(\mathbf{t},$ nat-to-v $(\mathbf{0},$ length $(a)), a)$

DEFINITION:
cvzbv-inc $(a) = $ cvzbv-v-adder $(\mathbf{t}, a,$ nat-to-v $(\mathbf{0},$ length $(a)))$

DEFINITION:
cvzbv-neg (a) = cvzbv-v-subtracter $(\mathbf{f}, a, \text{nat-to-v}(\mathbf{0}, \text{length}(a)))$

DEFINITION:
cvzbv-v-adder (c, a, b)
=
cvzbv (v-adder-carry-out (c, a, b),
     v-adder-overflowp (c, a, b),
     v-adder-output (c, a, b))

DEFINITION:
cvzbv-v-asr (a)
=
cvzbv (**if** listp (a) **then** nth $(\mathbf{0}, a)$
     else f endif,
     f,
     v-asr (a))

DEFINITION:
cvzbv-v-lsr (a)
=
cvzbv (**if** listp (a) **then** nth $(\mathbf{0}, a)$
     else f endif,
     f,
     v-lsr (a))

DEFINITION:
cvzbv-v-not (a) = cvzbv $(\mathbf{f}, \mathbf{f}, \text{v-not}(a))$

DEFINITION:
cvzbv-v-ror (c, a)
=
cvzbv (**if** nlistp (a) **then** c
     else nth $(\mathbf{0}, a)$ **endif**,
     f,
     v-ror (a, c))

DEFINITION:
cvzbv-v-subtracter (c, a, b)
=
cvzbv (v-subtracter-carry-out (c, a, b),
     v-subtracter-overflowp (c, a, b),
     v-subtracter-output (c, a, b))

DEFINITION:
firstn (*n*, *l*)
=
if listp (*l*)
 then if *n* ≅ 0 **then nil**
     else cons (car (*l*), firstn (*n* −1, cdr (*l*))) **endif**
 else nil endif

DEFINITION:
flags (*state*) = nth (**1**, *state*)

DEFINITION:
fm9001 (*state*, *n*)
=
if *n* ≅ 0 **then** *state*
 else fm9001 (fm9001-step (*state*, nat-to-v (**15**, reg-size)),*n* −1) **endif**

DEFINITION:
fm9001-alu-operation (*regs*, *flags*, *mem*, *ins*, *operand-a*, *operand-b*, *b-address*)
=
list (list (**if** store-resultp (store-cc (*ins*), *flags*) ∧ reg-direct-p (mode-b (*ins*))
       then write-mem (rn-b (*ins*),
                     regs,
                     bv (v-alu (c-flag (*flags*),
                            operand-a,
                            operand-b,
                            op-code (*ins*))))
         else *regs* **endif**,
         update-flags (*flags*,
                     set-flags (*ins*),
                     v-alu (c-flag (*flags*),
                         operand-a,
                         operand-b,
                         op-code (*ins*)))),
     if store-resultp (store-cc (*ins*), *flags*) ∧ (¬ reg-direct-p (mode-b (*ins*)))
      then write-mem (*b-address*,
                 mem,
                 bv (v-alu (c-flag (*flags*),
                         operand-a,
                         operand-b,
                         op-code (*ins*))))
       else *mem* **endif**)

DEFINITION:
fm9001-fetch (*regs*, *flags*, *mem*, *pc-reg*)
=

fm9001-operand-a (write-mem (*pc-reg*, *regs*, v-inc (read-mem (*pc-reg*, *regs*))),
        flags,
        mem,
        read-mem (read-mem (*pc-reg*, *regs*), *mem*))

DEFINITION:
fm9001-operand-a (*regs*, *flags*, *mem*, *ins*)
=

fm9001-operand-b (**if** a-immediate-p (*ins*) **then** *regs*
        elseif pre-dec-p (mode-a (*ins*))
        then write-mem (rn-a (*ins*),
                     regs,
                     v-dec (read-mem (rn-a (*ins*), *regs*)))
        elseif post-inc-p (mode-a (*ins*))
        then write-mem (rn-a (*ins*),
                     regs,
                     v-inc (read-mem (rn-a (*ins*), *regs*)))
        else *regs* **endif**,
        flags,
        mem,
        ins,
        if a-immediate-p (*ins*)
         then sign-extend (a-immediate (*ins*), **32**)
         elseif reg-direct-p (mode-a (*ins*))
         then read-mem (rn-a (*ins*), *regs*)
         elseif pre-dec-p (mode-a (*ins*))
         then read-mem (v-dec (read-mem (rn-a (*ins*), *regs*)), *mem*)
         else read-mem (read-mem (rn-a (*ins*), *regs*), *mem*) **endif**)

DEFINITION:
fm9001-operand-b (*regs*, *flags*, *mem*, *ins*, *operand-a*)
=
fm9001-alu-operation (**if** pre-dec-p (mode-b (*ins*))
                then write-mem (rn-b (*ins*),
                            regs,
                            v-dec (read-mem (rn-b (*ins*), *regs*)))
            elseif post-inc-p (mode-b (*ins*))
            then write-mem (rn-b (*ins*),
                          regs,
                          v-inc (read-mem (rn-b (*ins*), *regs*)))
        else *regs* **endif**,
        flags,
        mem,
        ins,
        operand-a,
        if reg-direct-p (mode-b (*ins*))
         then read-mem (rn-b (*ins*), *regs*)
         else read-mem (**if** pre-dec-p (mode-b (*ins*))
                       then v-dec (read-mem (rn-b (*ins*), *regs*))
                            else read-mem (rn-b (*ins*), *regs*) **endif**,
                      mem) **endif**,
        if pre-dec-p (mode-b (*ins*))
         then v-dec (read-mem (rn-b (*ins*), *regs*))
         else read-mem (rn-b (*ins*), *regs*) **endif**)

DEFINITION:
fm9001-step (*state*, *pc-reg*)
=
fm9001-fetch (regs (car (*state*)), flags (car (*state*)), cadr (*state*), *pc-reg*)

DEFINITION:
make-list (*n*, *value*)
=
if $n \cong 0$ **then nil**
 else cons (*value*, make-list (*n* −1, *value*)) **endif**

DEFINITION:
mode-a (*instruction*) = subrange (*instruction*, **4**, **5**)

DEFINITION:
mode-b (*instruction*) = subrange (*instruction*, **14**, **15**)

DEFINITION:
n (*cvzbv*) = v-negp (bv (*cvzbv*))

DEFINITION:
n-flag $(flags)$ = nth $(1, flags)$

DEFINITION:
n-set $(set\text{-}flags)$ = nth $(1, set\text{-}flags)$

DEFINITION:
nat-to-v (x, n)

=

if $n \cong 0$ then nil
 else cons $(\neg (x \bmod 2 \cong 0),$ nat-to-v $(x / 2, n - 1))$ endif

DEFINITION:
nth $(n, list)$

=

if $n \cong 0$ then car $(list)$
 else nth $(n - 1,$ cdr $(list))$ endif

DEFINITION:
op-code $(instruction)$ = subrange $(instruction, 24, 27)$

DEFINITION:
post-inc-p $(mode)$ = $mode$ = list (t, t)

DEFINITION:
pre-dec-p $(mode)$ = $mode$ = list (f, t)

SHELL DEFINITION:
Add the shell 'ram' of 1 argument, with
recognizer function symbol 'ramp' and
accessor 'ram-guts'.

DEFINITION:
read-mem $(v\text{-}addr, mem)$ = read-mem1 $(reverse (v\text{-}addr), mem)$

DEFINITION:
read-mem1 $(v\text{-}addr, mem)$

=

if stubp (mem) then stub-guts (mem)
 elseif nlistp $(v\text{-}addr)$
 then if ramp (mem) then ram-guts (mem)
       elseif romp (mem) then rom-guts (mem)
       else 0 endif
 elseif nlistp (mem) then 0
 elseif car $(v\text{-}addr)$ then read-mem1 $(cdr (v\text{-}addr),$ cdr $(mem))$
 else read-mem1 $(cdr (v\text{-}addr),$ car $(mem))$ endif

DEFINITION:
reg-direct-p (*mode*) = *mode* = list (**f**, **f**)

DEFINITION:
reg-size = **4**

DEFINITION:
regs (*state*) = nth (**0**, *state*)

DEFINITION:
rev1 (*x*, *sponge*)
=
if nlistp (*x*) **then** *sponge*
 else rev1 (cdr (*x*), cons (car (*x*), *sponge*)) **endif**

DEFINITION:
reverse (*x*) = rev1 (*x*, **nil**)

DEFINITION:
rn-a (*instruction*) = subrange (*instruction*, **0**, **3**)

DEFINITION:
rn-b (*instruction*) = subrange (*instruction*, **10**, **13**)

SHELL DEFINITION:
Add the shell 'rom' of 1 argument, with
recognizer function symbol 'romp' and
accessor 'rom-guts'.

DEFINITION:
set-flags (*instruction*) = subrange (*instruction*, **16**, **19**)

DEFINITION:
sign-extend (*v*, *n*)
=
if $n \cong 0$ **then nil**
 elseif nlistp (*v*) **then** make-list (*n*, **f**)
 elseif nlistp (cdr (*v*))
 then cons (boolfix (car (*v*)), make-list (*n* −1, boolfix (car (*v*))))
 else cons (boolfix (car (*v*)), sign-extend (cdr (*v*),*n* −1)) **endif**

DEFINITION:
store-cc (*instruction*) = subrange (*instruction*, **20**, **23**)

DEFINITION:
store-resultp (*store-cc*, *flags*)

=

if *store-cc* = list (**f, f, f, f**) then ¬ c-flag (*flags*)
 elseif *store-cc* = list (**t, f, f, f**) then c-flag (*flags*)
 elseif *store-cc* = list (**f, t, f, f**) then ¬ v-flag (*flags*)
 elseif *store-cc* = list (**t, t, f, f**) then v-flag (*flags*)
 elseif *store-cc* = list (**f, f, t, f**) then ¬ n-flag (*flags*)
 elseif *store-cc* = list (**t, f, t, f**) then n-flag (*flags*)
 elseif *store-cc* = list (**f, t, t, f**) then ¬ z-flag (*flags*)
 elseif *store-cc* = list (**t, t, t, f**) then z-flag (*flags*)
 elseif *store-cc* = list (**f, f, f, t**) then   (¬ c-flag (*flags*))
                                          ∧ (¬ z-flag (*flags*))
 elseif *store-cc* = list (**t, f, f, t**) then c-flag (*flags*) ∨ z-flag (*flags*)
 elseif *store-cc* = list (**f, t, f, t**)
 then   (n-flag (*flags*) ∧ v-flag (*flags*))
     ∨ ((¬ n-flag (*flags*)) ∧ (¬ v-flag (*flags*)))
 elseif *store-cc* = list (**t, t, f, t**)
 then   (n-flag (*flags*) ∧ (¬ v-flag (*flags*)))
     ∨ ((¬ n-flag (*flags*)) ∧ v-flag (*flags*))
 elseif *store-cc* = list (**f, f, t, t**)
 then   (n-flag (*flags*) ∧ v-flag (*flags*) ∧ (¬ z-flag (*flags*)))
     ∨ ((¬ n-flag (*flags*)) ∧ (¬ v-flag (*flags*)) ∧ (¬ z-flag (*flags*)))
 elseif *store-cc* = list (**t, f, t, t**)
 then   z-flag (*flags*)
     ∨ (n-flag (*flags*) ∧ (¬ v-flag (*flags*)))
     ∨ ((¬ n-flag (*flags*)) ∧ v-flag (*flags*))
 elseif *store-cc* = list (**f, t, t, t**) then t
 else f endif

SHELL DEFINITION:
Add the shell 'stub' of 1 argument, with
recognizer function symbol 'stubp' and
accessor 'stub-guts'.

DEFINITION:
subrange (*l*, *n*, *m*)

=

if $m < n$ then nil
 elseif $n \cong 0$
 then if $m \cong 0$ then list (car (*l*))
       else cons (car (*l*), subrange (cdr (*l*), 0, $m-1$)) endif
 else subrange (cdr (*l*), $n-1$, $m-1$) endif

DEFINITION:
update-flags (*flags*, *set-flags*, *cvzbv*)
=
list (b-if (z-set (*set-flags*), zb (*cvzbv*), z-flag (*flags*)),
    b-if (n-set (*set-flags*), n (*cvzbv*), n-flag (*flags*)),
    b-if (v-set (*set-flags*), v (*cvzbv*), v-flag (*flags*)),
    b-if (c-set (*set-flags*), c (*cvzbv*), c-flag (*flags*)))

DEFINITION:
v (*cvzbv*) = cadr (*cvzbv*)

DEFINITION:
v-adder (*c*, *a*, *b*)
=
if nlistp (*a*) **then** list (boolfix (*c*))
 else cons (b-xor3 (*c*, car (*a*), car (*b*)),
        v-adder (b-or (b-and (car (*a*), car (*b*)),
                b-or (b-and (car (*a*), *c*), b-and (car (*b*), *c*))),
          cdr (*a*),
          cdr (*b*))) **endif**

DEFINITION:
v-adder-carry-out (*c*, *a*, *b*) = nth (length (*a*), v-adder (*c*, *a*, *b*))

DEFINITION:
v-adder-output (*c*, *a*, *b*) = firstn (length (*a*), v-adder (*c*, *a*, *b*))

DEFINITION:
v-adder-overflowp (*c*, *a*, *b*)
=
b-and (b-equv (nth (length (*a*) -1, *a*), nth (length (*b*) -1, *b*)),
    b-xor (nth (length (*a*) -1, *a*), nth (length (*a*) -1, v-adder-output (*c*, *a*, *b*))))

DEFINITION:
v-alu (c, a, b, op)
=
if op = list (f, f, f, f) then cvzbv (f, f, v-buf (a))
 elseif op = list (t, f, f, f) then cvzbv-inc (a)
 elseif op = list (f, t, f, f) then cvzbv-v-adder (c, a, b)
 elseif op = list (t, t, f, f) then cvzbv-v-adder (f, a, b)
 elseif op = list (f, f, t, f) then cvzbv-neg (a)
 elseif op = list (t, f, t, f) then cvzbv-dec (a)
 elseif op = list (f, t, t, f) then cvzbv-v-subtracter (c, a, b)
 elseif op = list (t, t, t, f) then cvzbv-v-subtracter (f, a, b)
 elseif op = list (f, f, f, t) then cvzbv-v-ror (c, a)
 elseif op = list (t, f, f, t) then cvzbv-v-asr (a)
 elseif op = list (f, t, f, t) then cvzbv-v-lsr (a)
 elseif op = list (t, t, f, t) then cvzbv (f, f, v-xor (a, b))
 elseif op = list (f, f, t, t) then cvzbv (f, f, v-or (a, b))
 elseif op = list (t, f, t, t) then cvzbv (f, f, v-and (a, b))
 elseif op = list (f, t, t, t) then cvzbv-v-not (a)
 else cvzbv (f, f, v-buf (a)) endif

DEFINITION:
v-and (x, y)
=
if nlistp (x) then nil
 else cons (b-and (car (x), car (y)), v-and (cdr (x), cdr (y))) endif

DEFINITION:
v-asr (a) = v-shift-right $(a,$ nth (length (a) $-1, a$))

DEFINITION:
v-buf (x)
=
if nlistp (x) then nil
 else cons (b-buf (car (x)), v-buf (cdr (x))) endif

DEFINITION:
v-dec (x) = v-subtracter-output (t, nat-to-v (0, length (x)), x)

DEFINITION:
v-flag $(flags)$ = nth (2, $flags$)

DEFINITION:
v-inc (x) = v-adder-output (t, x, nat-to-v (0, length (x)))

DEFINITION:
v-lsr (a) = v-shift-right $(a,$ f)

DEFINITION:
v-negp (x)
=
if nlistp (x) **then f**
 elseif nlistp $(\mathrm{cdr}\,(x))$ **then** car (x)
 else v-negp $(\mathrm{cdr}\,(x))$ **endif**

DEFINITION:
v-not (x)
=
if nlistp (x) **then nil**
 else cons $(\mathrm{b\text{-}not}\,(\mathrm{car}\,(x)),\ \mathrm{v\text{-}not}\,(\mathrm{cdr}\,(x)))$ **endif**

DEFINITION:
v-nzerop (x)
=
if nlistp (x) **then f**
 else car $(x)\ \lor\ \mathrm{v\text{-}nzerop}\,(\mathrm{cdr}\,(x))$ **endif**

DEFINITION:
v-or (x, y)
=
if nlistp (x) **then nil**
 else cons $(\mathrm{b\text{-}or}\,(\mathrm{car}\,(x),\ \mathrm{car}\,(y)),\ \mathrm{v\text{-}or}\,(\mathrm{cdr}\,(x),\ \mathrm{cdr}\,(y)))$ **endif**

DEFINITION:
v-ror $(a, si) = $ v-shift-right (a, si)

DEFINITION:
v-set $(set\text{-}flags) = $ nth $(\mathbf{2},\ set\text{-}flags)$

DEFINITION:
v-shift-right (a, si)
=
if nlistp (a) **then nil**
 else append $(\mathrm{v\text{-}buf}\,(\mathrm{cdr}\,(a)),\ \mathrm{list}\,(\mathrm{boolfix}\,(si)))$ **endif**

DEFINITION:
v-subtracter-carry-out (c, a, b)
=
b-not $(\mathrm{v\text{-}adder\text{-}carry\text{-}out}\,(\mathrm{b\text{-}not}\,(c),\ \mathrm{v\text{-}not}\,(a),\ b))$

DEFINITION:
v-subtracter-output $(c, a, b) = $ v-adder-output $(\mathrm{b\text{-}not}\,(c),\ \mathrm{v\text{-}not}\,(a),\ b)$

DEFINITION:
v-subtracter-overflowp (c, a, b)
=
v-adder-overflowp (b-not (c), v-not (a), b)

DEFINITION:
v-xor (x, y)
=
if nlistp (x) **then nil**
 else cons (b-xor (car (x), car (y)), v-xor (cdr (x), cdr (y))) **endif**

DEFINITION:
v-zerop $(x) = \neg$ v-nzerop (x)

DEFINITION:
write-mem (v-$addr$, mem, $value$) = write-mem1 (reverse (v-$addr$), mem, $value$)

DEFINITION:
write-mem1 (v-$addr$, mem, $value$)
=
if stubp (mem) **then** mem
 elseif nlistp (v-$addr$)
 then if ramp (mem) **then** ram ($value$)
      else mem **endif**
 elseif nlistp (mem) **then** mem
 elseif car (v-$addr$)
 then cons (car (mem), write-mem1 (cdr (v-$addr$), cdr (mem), $value$))
 else cons (write-mem1 (cdr (v-$addr$), car (mem), $value$), cdr (mem)) **endif**

DEFINITION:
z-flag ($flags$) = nth (0, $flags$)

DEFINITION:
z-set (set-$flags$) = nth (0, set-$flags$)

DEFINITION:
zb ($cvzbv$) = caddr ($cvzbv$)

Appendix IV

The Formal Implementation

In this appendix we formally present the implementation of Piton on the FM9001. As before, we divide the appendix into two parts: a guide to the formal listing and then the listing itself.

IV.1. A Guide to the Formal Implementation

The FM9001 implementation of Piton is expressed in the function 'load'.

DEFINITION:
load (*p, boot-lst, load-addr*)
=
m⇒fm9001 (i⇒m (r⇒i (p⇒r (*p*)), *boot-lst, load-addr*))

We discuss each phase in turn.

IV.1.1. The Formal Definition of Resource Representation

R-states are represented by the new shell class

SHELL DEFINITION:
Add the shell 'r-state' of 15 arguments, with
recognizer function symbol 'r-statep', and
accessors 'r-pc', 'r-cfp', 'r-csp', 'r-tsp', 'r-x', 'r-y',
        'r-c-flg', 'r-v-flg', 'r-n-flg', 'r-z-flg',
           'r-prog-segment', 'r-usr-data-segment', 'r-sys-data-segment',
           'r-word-size', and 'r-psw'.

The r-state corresponding to a given p-state has exactly the same program counter, program segment, data segment (here called "user data segment"), word size and psw. However, r-states also have five additional "registers" ('r-cfp', 'r-csp', 'r-tsp', 'r-x' and 'r-y'), four "flag registers," and a "system data segment." These components of an r-state specify how the corresponding resources of the underlying FM9001 machine will be used to implement the resources of the Piton machine.

The r-state corresponding to a given p-state is produced by the function 'p⇒r', which implements the resource representation phase of 'load'.

DEFINITION:
p⇒r (p)
=
r-state (p-pc (p),
      p⇒r_cfp (p-ctrl-stk (p), p-max-ctrl-stk-size (p)),
      p⇒r_csp (p-ctrl-stk (p), p-max-ctrl-stk-size (p)),
      p⇒r_tsp (p-temp-stk (p), p-max-temp-stk-size (p)),
      `'(nat 0)`,
      `'(nat 0)`,
      `'(bool f)`,
      `'(bool f)`,
      `'(bool f)`,
      `'(bool f)`,
      p-prog-segment (p),
      p-data-segment (p),
      p⇒r_sys-data-segment (p-ctrl-stk (p),
                                p-max-ctrl-stk-size (p),
                                  p-temp-stk (p),
                                  p-max-temp-stk-size (p)),
      p-word-size (p),
      p-psw (p))

Observe that the program counter, program segment, data segment, word size and psw of the given p-state are copied as is into the r-state. The three stack registers are initialized by the three functions 'p⇒r_cfp', 'p⇒r_csp', and 'p⇒r_tsp'. The two temporary registers are initialized to the natural number 0 and the four flags are all initially **f**. The system data segment of the r-state is determined by 'p⇒r_sys-data-segment'.

We will look in detail at the handling of the temporary stack, since it is the simpler of the two stacks. The initial stack pointer, 'r-tsp', is computed by

DEFINITION:
p⇒r_tsp (*stk*, *max*)
=
tag (`'sys-addr`, cons (`'tstk`, *max* − length (*stk*)))

which takes as its arguments the actual Piton temporary stack, *stk*, and the maximum size of that stack, *max*. The function yields a system data address into the **tstk** area. The arithmetic accounts for the fact that the stack elements are loaded into the high end of the **tstk** area.

The system data segment itself is constructed by

DEFINITION:
p⟹r_sys-data-segment (*ctrl-stk, max-ctrl-stk-size, temp-stk, max-temp-stk-size*)
=
list (p⟹r_ctrl-stk (*ctrl-stk, max-ctrl-stk-size*),
    p⟹r_temp-stk (*temp-stk, max-temp-stk-size*),
    list (**'full-ctrl-stk-addr**,
        tag (**'sys-addr, '(cstk . 0)**))),
    list (**'full-temp-stk-addr**,
        tag (**'sys-addr, '(tstk . 0)**))),
    list (**'empty-temp-stk-addr**,
        tag (**'sys-addr**, cons (**'tstk**, *max-temp-stk-size*)))))

which describes a segment with five areas. The second one is the temporary stack area, constructed by

DEFINITION:
p⟹r_temp-stk (*temp-stk, max-temp-stk-size*)
=
cons (**'tstk**,
    append (nat-0s (*max-temp-stk-size* − length (*temp-stk*)),
        append (*temp-stk*, list (tag (**'nat, 0**)))))

which takes the Piton temporary stack and its maximum size and yields an area named **tstk** associated with an array of length *max-temp-stk-size* + 1. The array is initialized with the temporary stack from the p-state, padded at the high end of the area by a single natural number **0** and padded on the low end by as many 0s as necessary to make the array the correct length.

The construction of the control stack area is much more subtle but easily followed from the informal discussion.

IV.1.2. *The Formal Definition of the Compiler*

The compilation phase translates Piton to i-code. The state transformation associated with this is implemented by 'r⟹i', which converts an r-state into an i-state, where

SHELL DEFINITION:
Add the shell 'i-state' of 15 arguments, with
recognizer function symbol 'i-statep', and
accessors 'i-pc', 'i-cfp', 'i-csp', 'i-tsp', 'i-x', 'i-y',
        'i-c-flg', 'i-v-flg', 'i-n-flg', 'i-z-flg',
        'i-prog-segment', 'i-usr-data-segment', 'i-sys-data-segment',
        'i-word-size', and 'i-psw'.

The i-state corresponding to a given r-state is produced by 'r⟹i', below. The i-state is identical in all components except the program counter, the program segment, and the psw.

DEFINITION:
r⇒i (r)
=
i-state (r⇒i_pc (r-pc (r), r-prog-segment (r)),
        r-cfp (r),
        r-csp (r),
        r-tsp (r),
        r-x (r),
        r-y (r),
        r-c-flg (r),
        r-v-flg (r),
        r-n-flg (r),
        r-z-flg (r),
        icompile (r-prog-segment (r)),
        r-usr-data-segment (r),
        r-sys-data-segment (r),
        r-word-size (r),
        r⇒i_psw (r-psw (r)))

The code in the i-state program segment is produced by compiling the Piton code in
the r-state program segment. The program counter of the i-state is arranged to point
to the beginning of the i-code block for the current instruction in the r-state. The psw
of the i-state is either **'run** or an error— **'halt** is coerced to **'run**.

The compiler is 'icompile'.

DEFINITION:
icompile (*programs*)

=

if nlistp (*programs*) **then nil**
 else cons (icompile-program (car (*programs*)), icompile (cdr (*programs*))) **endif**

This function maps over the list of programs and compiles each with 'icompile-
program'.

DEFINITION:
icompile-program (*program*)

=

cons (name (*program*),
      append (generate-prelude (*program*),
              append (icompile-program-body (program-body (*program*),
                                             0, *program*),
                      generate-postlude (*program*))))

'icompile-program' conses the program name onto the result of compiling the body
of the program and sandwiching it between the prelude and the postlude. The cons
pair returned becomes the definition of a program area in the program segment of the
i-state.

The prelude is generated with 'generate-prelude',

DEFINITION:
generate-prelude (*program*)

=

append (cons (dl (cons (name (*program*), ' (**prelude**)),
                    ' (**prelude**),
                    ' (**cpush_cfp**)),
                ' ((**move_cfp_csp**))),
            append (generate-prelude1 (reverse (temp-var-dcls (*program*))),
                    generate-prelude2 (formal-vars (*program*))))

Note that the basic block generated by 'generate-prelude' begins with a def-label form defining a label of the form (*name* **PRELUDE**). The first instruction is the **cpush_cfp** that pushes the old cfp word of the new frame. The next instruction saves **csp** in **cfp**. Then we lay down the code to initialize the temporary variables (generated by 'generate-prelude1') and the formals ('generate-prelude2').

Below is 'generate-prelude1'. It generates a list of **cpush_*** instructions, each of which is followed by the initial value specified in the temporary variable declaration of the program being compiled.

DEFINITION:
generate-prelude1 (*temp-var-dcls*)

=

if nlistp (*temp-var-dcls*) then nil
else cons (' (**cpush_***),
        cons (cadar (*temp-var-dcls*),
            generate-prelude1 (cdr (*temp-var-dcls*)))) endif

'Generate-prelude2' generates a list of **cpush_<tsp>+** instructions as long as the formals of the program. These instructions move the actual parameters from the temporary stack to the control stack.

DEFINITION:
generate-prelude2 (*formal-vars*)

=

if nlistp (*formal-vars*) then nil
else cons (' (**cpush_<tsp>+**), generate-prelude2 (cdr (*formal-vars*))) endif

The generation of the postlude is similar in spirit to the examples given here.
The body of each program is compiled with

DEFINITION:
icompile-program-body (*lst, pcn, program*)

=

if nlistp (*lst*) then nil
else append (icode (car (*lst*), *pcn, program*),
        icompile-program-body (cdr (*lst*), 1+ *pcn, program*)) endif

The function 'icode' takes a Piton instruction, together with the offset (*pcn*) at which it occurs in the body and the entire program (*program*) containing the instruction, and generates the basic block of i-code for the given instruction. 'icode' works by using 'icode1' to generate the i-code and then attaching a label to that block to mark the location at which the corresponding Piton **pc** begins. (The definition of 'icode' is on page 273). 'Icode1' is simply another big case statement on the opcode of the Piton instruction.

Each Piton instruction has an i-code generator for its basic block. Our naming convention is that the code for the Piton instruction with name *opcode* is generated by the function named 'icode-*opcode*'.

Below we show the generator for the **push-constant** instruction, i.e., 'icode-push-constant'.

DEFINITION:
icode-push-constant (*ins*, *pcn*, *program*)

=

list (' **(tpush_*)**,
    if cadr (*ins*) = ' **pc then** tag (' **pc**, cons (name (*program*), 1+ *pcn*))
      elseif nlistp (cadr (*ins*)) **then** pc (cadr (*ins*), *program*)
      else cadr (*ins*) **endif)**

Observe that the i-code generated for **push-constant** contains two items. The first is a **tpush_*** instruction. The second is the Piton object to be used as immediate data for the **tpush_***.

A more interesting instruction, perhaps, is **push-local**.

DEFINITION:
icode-push-local (*ins*, *pcn*, *program*)

=

list (' **(move_x_*)**,
    tag (' **nat**, offset-from-csp (cadr (*ins*), *program*)),
    ' **(add_x{n}_csp)**,
    ' **(tpush_<x{s}>))**

This function generates a list of four i-code instructions and data, using 'offset-from-csp' to determine the position within the local variables of the local variable to be pushed.

This completes our tour through the compilation phase. The reader is invited to study the i-code generators for each of the Piton instructions.

IV.1.3. The Formal Definition of the Link-Assembler

The link-assembler takes an i-state, containing symbolic programs and data, and produces an "m-state" in which programs and data are represented as contiguous sequences of bit vectors. An m-state is a structure very similar to an FM9001 state, except that the registers and memory are represented as linear lists of bit vectors instead of as ram trees.

SHELL DEFINITION:
Add the shell 'm-state' of 6 arguments, with
recognizer function symbol 'm-statep', and
accessors 'm-regs',
        'm-c-flg', 'm-v-flg', 'm-n-flg', 'm-z-flg',
        'm-mem'.

The link-assembler converts an i-state to an m-state by replacing all the i-code instructions in the program segment by bit vectors and all the symbolic data objects (in the stacks, programs, registers and data) by bit vectors. The implementation of the link-assembler is the function

DEFINITION:
i⇒m (*i, boot-lst, load-addr*)
=
let *tables* **be** i-link-tables (*i, load-addr*),
   w **be** i-word-size (*i*) **in**
m-state (list (link-word (**' (nat 0)**, *tables, w*),
           link-word (i-cfp (*i*), *tables, w*),
           link-word (i-csp (*i*), *tables, w*),
           link-word (i-tsp (*i*), *tables, w*),
           link-word (i-x (*i*), *tables, w*),
           link-word (i-y (*i*), *tables, w*),
           link-word (**' (nat 0)**, *tables, w*),
           link-word (**' (nat 0)**, *tables, w*),
           link-word (**' (nat 0)**, *tables, w*),
           link-word (**' (nat 0)**, *tables, w*),
           link-word (**' (nat 0)**, *tables, w*),
           link-word (**' (nat 0)**, *tables, w*),
           link-word (**' (nat 0)**, *tables, w*),
           link-word (**' (nat 0)**, *tables, w*),
           link-word (i-pc (*i*), *tables, w*)),
       bool-to-logical (untag (i-c-flg (*i*))),
       bool-to-logical (untag (i-v-flg (*i*))),
       bool-to-logical (untag (i-n-flg (*i*))),
       bool-to-logical (untag (i-z-flg (*i*))),
       link-mem (*boot-lst*,
               load-addr,
               i-usr-data-segment (*i*),
               i-prog-segment (*i*),
               i-sys-data-segment (*i*),
               tables,
               w)) **endlet**

The 'list-expression' in the first argument of 'm-state' above describes the 16 registers. Note that 'i-cfp' is allocated to register 1, 'i-csp' to register 2, 'i-tsp' to register 3, and 'i-x' and 'i-y' to registers 4 and 5. The 'i-pc' is allocated to register 15. The other registers are unused. Next come the four condition code registers, which are mapped from Piton Booleans, **t** and **f**, to the Booleans of the Nqthm (and the FM9001 model) **t** and **f**. Finally, comes the memory created by 'link-mem'. The memory is produced by mapping over the boot code (as specified by the *boot-lst* and *load-addr*), the user data segment, the i-code program segment, and the system data segment and appending together the results of linking every word with the function 'link-word'. Before this is done the link tables are computed by 'i-link-tables'.

'link-word' operates by first determining whether the word is an i-code instruction or a data word and using the appropriate linker.

DEFINITION:
link-word (*x*, *link-tables*, *word-size*)

=

if icode-instructionp (*x*) **then** link-instr-word (*x*, *word-size*)
 else link-data-word (*x*, *link-tables*, *word-size*) **endif**

   I-code instructions are linked with

DEFINITION:
link-instr-word (*ins*, *word-size*)

=

mci (cadr (assoc (car (*ins*), link-instruction-alist)), *word-size*)

The 'assoc' above looks up the i-code opcode in the table ('link-instruction-alist')
that expands it into an assembly instruction. Then 'mci' (*machine code instruction*)
assembles the instruction into a bit vector.

DEFINITION:
mci (*ins*, *word-size*)

=

pack-instruction (extract-op (car (*ins*)),
                    extract-move-bits (car (*ins*)),
                    extract-cvnz (cadr (*ins*)),
                    extract-mode (caddr (*ins*)),
                    extract-reg (caddr (*ins*)),
                    extract-mode (cadddr (*ins*)),
                    extract-reg (cadddr (*ins*)),
                    word-size)

   If, on the other hand, the word to be linked is a data word, 'link-data-word' is
used. It case splits on the type of the data word and uses the appropriate mapping to
bit vectors.

DEFINITION:
link-data-word (*x*, *link-tables*, *word-size*)

=

case on type (*x*):
 case = **nat** **then** nat-to-v (untag (*x*), *word-size*)
 case = **int** **then** int-to-v (untag (*x*), *word-size*)
 case = **bitv** **then** bitv-to-v (untag (*x*), *word-size*)
 case = **bool** **then** bool-to-v (untag (*x*), *word-size*)
 case = **addr** **then** addr-to-v (untag (*x*), usr-data-links (*link-tables*), *word-size*)
 case = **subr** **then** subr-to-v (untag (*x*), prog-links (*link-tables*), *word-size*)
 case = **sys-addr**
  then sys-addr-to-v (untag (*x*), sys-data-links (*link-tables*), *word-size*)
 case = **pc** **then** label-to-v (untag (*x*), prog-label-tables (*link-tables*), *word-size*)
 case = **ipc** **then** ipc-to-v (untag (*x*), prog-links (*link-tables*), *word-size*)
 otherwise nat-to-v (0, *word-size*) **endcase**

IV.1.4. The Formal Definition of the Image Constructor

We are finally ready to produce an FM9001 state. This is done by 'm⇒fm9001'.

DEFINITION:
m⇒fm9001 (*m*)
=
list (list (ram-tree (m-regs (*m*), **16**),
          list (m-z-flg (*m*), m-n-flg (*m*), m-v-flg (*m*), m-c-flg (*m*))),
     ram-tree (m-mem (*m*), **2³²**))

Observe that the registers and memory of the FM9001 state are created by the function 'ram-tree' from the corresponding linear lists in the m-state. (Our informal description of ram trees parameterized them by their depth but they are formally parameterized by their size.)

Here is the definition of 'ram-tree'.

DEFINITION:
ram-tree (*lst, size*)
=
if *size* ≅ **0** **then** stub (nat-to-v (**0, 32**))
 elseif nlistp (*lst*) **then** stub (nat-to-v (**0, 32**))
 elseif *size* = **1** **then** ram (car (*lst*))
 else cons (ram-tree (firstn (*size* / **2**, *lst*), *size* / **2**),
              ram-tree (restn (*size* / **2**, *lst*), *size* / **2**)) **endif**

This completes our tour through the link-assembler. The reader is urged to read the code for assembling instructions and mapping each type of data object into bit vectors.

Reader uninterested in pursuing the formal definition of the implementation should skip to page 299.

IV.2. Alphabetical Listing of the Implementation

DEFINITION:
absolute-address (*adp, link-table*)
=
base-address (adp-name (*adp*), *link-table*) + adp-offset (*adp*)

DEFINITION:
addr-to-v (*adp, usr-data-links, word-size*)
=
nat-to-v (absolute-address (*adp, usr-data-links*), *word-size*)

DEFINITION:
base-address (*name, link-table*) = cdr (assoc (*name, link-table*))

DEFINITION:
bitv-to-v (*lst*, *word-size*)

=

if *word-size* ≅ **0** **then nil**
 else append (bitv-to-v (cdr (*lst*),*word-size* −1),
           list (**if** car (*lst*) = **0** **then f**
                else t endif)) **endif**

DEFINITION:
bool-to-logical (*b*)

=

if *b* = ' **f** **then f**
 else t endif

DEFINITION:
bool-to-v (*b*, *word-size*)

=

if *b* = ' **f** **then** nat-to-v (**0**, *word-size*)
 else nat-to-v (**1**, *word-size*) **endif**

DEFINITION:
boot-code (*lst*, *n*, *word-size*)

=

if *n* ≅ **0** **then nil**
 else cons (nat-to-v (car (*lst*), *word-size*),
          boot-code (cdr (*lst*),*n* −1, *word-size*)) **endif**

DEFINITION:
dl (*lab*, *comment*, *ins*) = list (' **dl**, *lab*, *comment*, *ins*)

DEFINITION:
dl-block (*lab*, *comment*, *block*)

=

cons (dl (*lab*, *comment*, car (*block*)), cdr (*block*))

DEFINITION:
extract-cvnz (*flg-names*)
=

$\quad$ ($\quad$ **if** '**c** $\in$ *flg-names* **then 1**
$\qquad$ **else 0 endif**
$\quad \times$ 2^3)
$+$ ($\quad$ **if** '**v** $\in$ *flg-names* **then 1**
$\qquad$ **else 0 endif**
$\quad \times$ 2^2)
$+$ ($\quad$ **if** '**n** $\in$ *flg-names* **then 1**
$\qquad$ **else 0 endif**
$\quad \times$ 2^1)
$+$ ($\quad$ **if** '**z** $\in$ *flg-names* **then 1**
$\qquad$ **else 0 endif**
$\quad \times$ 2^0)

DEFINITION:
extract-mode (*reg-spec*)
=

if litatom (*reg-spec*) **then 0**
$\quad$ **elseif** cdr (*reg-spec*) = **nil then 1**
$\quad$ **elseif** car (*reg-spec*) = **-1 then 2**
$\quad$ **else 3 endif**

DEFINITION:
extract-move-bits (*opcode*)
=

case on *opcode*:
$\quad$ **case = move then 14**
$\quad$ **case = move-nc then 0**
$\quad$ **case = move-c then 1**
$\quad$ **case = move-nv then 2**
$\quad$ **case = move-v then 3**
$\quad$ **case = move-nz then 6**
$\quad$ **case = move-z then 7**
$\quad$ **case = move-nn then 4**
$\quad$ **case = move-n then 5**
$\quad$ **otherwise 14 endcase**

DEFINITION:
extract-op (*opcode*)
=
cadr (assoc (*opcode*,
           '((incr 1)
             (addc 2)
             (add 3)
             (neg 4)
             (decr 5)
             (subb 6)
             (sub 7)
             (ror 8)
             (asr 9)
             (lsr 10)
             (xor 11)
             (or 12)
             (and 13)
             (not 14)
             (move 15)
             (move-nc 15)
             (move-c 15)
             (move-nv 15)
             (move-v 15)
             (move-nz 15)
             (move-z 15)
             (move-nn 15)
             (move-n 15))))

DEFINITION:
extract-reg (*reg-spec*)
=
cadr (assoc (extract-reg1 (*reg-spec*),
           '((pc 15)
            (cfp 1)
            (csp 2)
            (tsp 3)
            (x 4)
            (y 5))))

DEFINITION:
extract-reg1 (*reg-spec*)
=
if litatom (*reg-spec*) then *reg-spec*
 elseif cdr (*reg-spec*) = nil then car (*reg-spec*)
 elseif car (*reg-spec*) = -1 then cadr (*reg-spec*)
 else car (*reg-spec*) endif

DEFINITION:
find-position-of-var (*var*, *lst*)

=

if nlistp (*lst*) **then** 0
 elseif *var* = car (*lst*) **then** 0
 else 1+ find-position-of-var (*var*, cdr (*lst*)) **endif**

DEFINITION:
generate-postlude (*program*)

=

cons (dl (cons (name (*program*), length (program-body (*program*)))),
   ' (postlude),
   ' (move_csp_cfp)),
  ' ((cpop_cfp) (cpop_pc)))

DEFINITION:
generate-prelude (*program*)

=

append (cons (dl (cons (name (*program*), **' (prelude)**)),
     ' (prelude),
     ' (cpush_cfp)),
    ' ((move_cfp_csp))),
   append (generate-prelude1 (reverse (temp-var-dcls (*program*))),
     generate-prelude2 (formal-vars (*program*))))

DEFINITION:
generate-prelude1 (*temp-var-dcls*)

=

if nlistp (*temp-var-dcls*) **then nil**
 else cons (**' (cpush_*)**,
   cons (cadar (*temp-var-dcls*),
    generate-prelude1 (cdr (*temp-var-dcls*)))) **endif**

DEFINITION:
generate-prelude2 (*formal-vars*)

=

if nlistp (*formal-vars*) **then nil**
 else cons (**' (cpush_<tsp>+)**, generate-prelude2 (cdr (*formal-vars*))) **endif**

DEFINITION:

i⇒m (*i, boot-lst, load-addr*)

=

let *tables* **be** i-link-tables (*i, load-addr*),
    w **be** i-word-size (*i*) **in**
m-state (list (link-word (' **(nat 0)**, *tables, w*),
            link-word (i-cfp (*i*), *tables, w*),
            link-word (i-csp (*i*), *tables, w*),
            link-word (i-tsp (*i*), *tables, w*),
            link-word (i-x (*i*), *tables, w*),
            link-word (i-y (*i*), *tables, w*),
            link-word (' **(nat 0)**, *tables, w*),
            link-word (' **(nat 0)**, *tables, w*),
            link-word (' **(nat 0)**, *tables, w*),
            link-word (' **(nat 0)**, *tables, w*),
            link-word (' **(nat 0)**, *tables, w*),
            link-word (' **(nat 0)**, *tables, w*),
            link-word (' **(nat 0)**, *tables, w*),
            link-word (' **(nat 0)**, *tables, w*),
            link-word (' **(nat 0)**, *tables, w*),
            link-word (i-pc (*i*), *tables, w*)),
        bool-to-logical (untag (i-c-flg (*i*))),
        bool-to-logical (untag (i-v-flg (*i*))),
        bool-to-logical (untag (i-n-flg (*i*))),
        bool-to-logical (untag (i-z-flg (*i*))),
        link-mem (*boot-lst,*
                load-addr,
                i-usr-data-segment (*i*),
                i-prog-segment (*i*),
                i-sys-data-segment (*i*),
                tables,
                w)) **endlet**

DEFINITION:

i-link-tables (*i, load-addr*)

=

list (link-table-for-segment (i-prog-segment (*i*),
                load-addr
            +
                segment-length (i-usr-data-segment (*i*))),
    link-table-for-prog-labels (i-prog-segment (*i*),
                load-addr
            +
                segment-length (i-usr-data-segment (*i*))),
    link-table-for-segment (i-usr-data-segment (*i*), *load-addr*),
    link-table-for-segment (i-sys-data-segment (*i*),
                load-addr
            + segment-length (i-prog-segment (*i*))
            + segment-length (i-usr-data-segment (*i*)))))

SHELL DEFINITION:
Add the shell 'i-state' of 15 arguments, with
recognizer function symbol 'i-statep', and
accessors 'i-pc', 'i-cfp', 'i-csp', 'i-tsp', 'i-x', 'i-y',
           'i-c-flg', 'i-v-flg', 'i-n-flg', 'i-z-flg',
           'i-prog-segment', 'i-usr-data-segment', 'i-sys-data-segment',
           'i-word-size', and 'i-psw'.

DEFINITION:
icode (*ins, pcn, program*)

=

dl-block (cons (name (*program*), *pcn*), *ins*, icode1 (unlabel (*ins*), *pcn*, *program*))

DEFINITION:
icode-add-addr (*ins, pcn, program*)

=

```
'((tpop_x)  (add_<tsp>{a}_x{n}))
```


DEFINITION:
icode-add-int (*ins, pcn, program*)

=

```
'((tpop_x)  (add_<tsp>{i}_x{i}))
```


DEFINITION:
icode-add-int-with-carry (*ins, pcn, program*)

=

```
'((tpop_x)
  (tpop_y)
  (asr_<c>_<tsp>_<tsp>{b})
  (addc_<v>_x{i}_y{i})
  (move-v_<tsp>_*)
  (bool t)
  (tpush_x))
```


DEFINITION:
icode-add-nat (*ins, pcn, program*)

=

```
'((tpop_x)  (add_<tsp>{n}_x{n}))
```

DEFINITION:
icode-add-nat-with-carry (*ins*, *pcn*, *program*)
=
```
'((tpop_x)
  (tpop_y)
  (asr_<c>_<tsp>_<tsp>{b})
  (addc_<c>_x{n}_y{n})
  (move-c_<tsp>_*)
  (bool t)
  (tpush_x))
```

DEFINITION:
icode-add1-int (*ins*, *pcn*, *program*)
=
```
'((incr_<tsp>_<tsp>{i}))
```

DEFINITION:
icode-add1-nat (*ins*, *pcn*, *program*)
=
```
'((incr_<tsp>_<tsp>{n}))
```

DEFINITION:
icode-and-bitv (*ins*, *pcn*, *program*)
=
```
'((tpop_x) (and_<tsp>{v}_x{v}))
```

DEFINITION:
icode-and-bool (*ins*, *pcn*, *program*)
=
```
'((tpop_x) (and_<tsp>{b}_x{b}))
```

DEFINITION:
icode-call (*ins*, *pcn*, *program*)
=
list (' (cpush_*),
    tag ('pc, cons (name (*program*), 1+ *pcn*)),
    '(jump_*),
    tag ('pc, cons (cadr (*ins*), ' (prelude))))

DEFINITION:
icode-deposit (*ins*, *pcn*, *program*)
=
```
'((tpop_x) (tpop_<x{a}>))
```

DEFINITION:
icode-deposit-temp-stk (*ins*, *pcn*, *program*)
=
```
'((tpop_y)
  (incr_y_y{n})
  (move_x_*)
  (sys-addr (empty-temp-stk-addr . 0))
  (move_x_<x{s}>)
  (sub_x{s}_y{n})
  (tpop_<x{s}>))
```

DEFINITION:
icode-div2-nat (*ins*, *pcn*, *program*)
=
```
'((tpop_<c>_x)
  (lsr_<c>_x_x{n})
  (tpush_x)
  (tpush_*)
  (nat 0)
  (move-c_<tsp>_*)
  (nat 1))
```

DEFINITION:
icode-eq (*ins*, *pcn*, *program*)
=
```
'((tpop_x)
  (xor_<z>_<tsp>_x)
  (xor_<tsp>_<tsp>)
  (move-z_<tsp>_*)
  (bool t))
```

DEFINITION:
icode-fetch (*ins*, *pcn*, *program*)
=
```
'((tpop_x) (tpush_<x{a}>))
```

DEFINITION:
icode-fetch-temp-stk (*ins*, *pcn*, *program*)
=
```
'((tpop_y)
  (incr_y_y{n})
  (move_x_*)
  (sys-addr (empty-temp-stk-addr . 0))
  (move_x_<x{s}>)
  (sub_x{s}_y{n})
  (tpush_<x{s}>))
```

DEFINITION:
icode-instructionp (*ins*) = cdr (*ins*) = **nil**

DEFINITION:
icode-int-to-nat (*ins*, *pcn*, *program*) = ' ((**int-to-nat**))

DEFINITION:
icode-jump (*ins*, *pcn*, *program*)
=
list (' (**jump_***), pc (cadr (*ins*), *program*))

DEFINITION:
icode-jump-case (*ins*, *pcn*, *program*)
=
append (' ((**tpop_x**)
       (**add_x_x{n}**)
       (**add_pc_x{n}**))),
    jump_*-lst (cdr (*ins*), *program*))

DEFINITION:
icode-jump-if-temp-stk-empty (*ins*, *pcn*, *program*)
=
list (' (**move_y_tsp**),
  ' (**move_x_***),
  ' (**sys-addr (empty-temp-stk-addr . 0)**),
  ' (**move_x_<x{s}>**),
  ' (**sub_<z>_x{s}_y{s}**),
  ' (**move_x_***),
  pc (cadr (*ins*), *program*),
  ' (**jump-z_x**))

DEFINITION:
icode-jump-if-temp-stk-full (*ins*, *pcn*, *program*)
=
list (' (**move_x_tsp**),
  ' (**move_y_***),
  ' (**sys-addr (full-temp-stk-addr . 0)**),
  ' (**move_y_<y{s}>**),
  ' (**sub_<z>_x{s}_y{s}**),
  ' (**move_x_***),
  pc (cadr (*ins*), *program*),
  ' (**jump-z_x**))

DEFINITION:
icode-locn (*ins*, *pcn*, *program*)
=
list (' (move_x_*),
    tag ('nat, offset-from-csp (cadr (*ins*), *program*)),
   ' (add_x{n}_csp),
   ' (move_x_<x{s}>),
   ' (add_x{n}_csp),
   ' (tpush_<x{s}>))

DEFINITION:
icode-lsh-bitv (*ins*, *pcn*, *program*)
=
' ((add_<tsp>_<tsp>{v}))

DEFINITION:
icode-lt-addr (*ins*, *pcn*, *program*)
=
' ((tpop_x)
  (sub_<c>_<tsp>{a}_x{a})
  (xor_<tsp>_<tsp>)
  (move-c_<tsp>_*)
  (bool t))

DEFINITION:
icode-lt-int (*ins*, *pcn*, *program*)
=
' ((tpop_x)
  (sub_<nv>_<tsp>{i}_x{i})
  (move_<tsp>_*)
  (bool f)
  (move-v_<tsp>_*)
  (bool t)
  (move_x_*)
  (bool f)
  (move-n_x_*)
  (bool t)
  (xor_<tsp>{b}_x{b}))

DEFINITION:
icode-lt-nat (*ins*, *pcn*, *program*)
=
' ((tpop_x)
  (sub_<c>_<tsp>{n}_x{n})
  (xor_<tsp>_<tsp>)
  (move-c_<tsp>_*)
  (bool t))

DEFINITION:
icode-mult2-nat (*ins*, *pcn*, *program*)
=
```
'((add_<tsp>_<tsp>{n}))
```


DEFINITION:
icode-mult2-nat-with-carry-out (*ins*, *pcn*, *program*)
=
```
'((tpop_x)
  (add_<c>_x_x{n})
  (tpush_*)
  (bool f)
  (move-c_<tsp>_*)
  (bool t)
  (tpush_x))
```


DEFINITION:
icode-neg-int (*ins*, *pcn*, *program*)
=
```
'((neg_<tsp>_<tsp>{i}))
```


DEFINITION:
icode-no-op (*ins*, *pcn*, *program*) = `'((move_x_x))`

DEFINITION:
icode-not-bitv (*ins*, *pcn*, *program*)
=
```
'((not_<tsp>_<tsp>{v}))
```


DEFINITION:
icode-not-bool (*ins*, *pcn*, *program*)
=
```
'((xor_<tsp>{b}_*{b}) (bool t))
```


DEFINITION:
icode-or-bitv (*ins*, *pcn*, *program*)
=
```
'((tpop_x) (or_<tsp>{v}_x{v}))
```


DEFINITION:
icode-or-bool (*ins*, *pcn*, *program*)
=
```
'((tpop_x) (or_<tsp>{b}_x{b}))
```


DEFINITION:
icode-pop (*ins*, *pcn*, *program*) = `'((tpop_x))`

DEFINITION:
icode-pop* (*ins*, *pcn*, *program*)
=
list(' (add_tsp_*{n}), tag('nat, cadr(*ins*)))

DEFINITION:
icode-pop-call (*ins*, *pcn*, *program*)
=
list(' (tpop_x),
   ' (cpush_*),
   tag('pc, cons (name (*program*), 1+ *pcn*)),
   ' (jump_x{subr}))

DEFINITION:
icode-pop-global (*ins*, *pcn*, *program*)
=
list(' (move_x_*),
   tag('addr, cons (cadr (*ins*), 0)),
   ' (tpop_<x{a}>))

DEFINITION:
icode-pop-local (*ins*, *pcn*, *program*)
=
list(' (move_x_*),
   tag('nat, offset-from-csp (cadr (*ins*), *program*)),
   ' (add_x{n}_csp),
   ' (tpop_<x{s}>))

DEFINITION:
icode-pop-locn (*ins*, *pcn*, *program*)
=
list(' (move_x_*),
   tag('nat, offset-from-csp (cadr (*ins*), *program*)),
   ' (add_x{n}_csp),
   ' (move_x_<x{s}>),
   ' (add_x{n}_csp),
   ' (tpop_<x{s}>))

DEFINITION:
icode-popj (*ins*, *pcn*, *program*) = ' ((tpop_pc))

DEFINITION:
icode-popn (*ins*, *pcn*, *program*)
=
' ((tpop_x) (add_tsp_x{n}))

DEFINITION:
icode-push-constant (*ins*, *pcn*, *program*)

=

list(`' (tpush_*)`,
    if cadr (*ins*) = `'pc` **then** tag (`'pc`, cons (name (*program*), 1+ *pcn*))
    elseif nlistp (cadr (*ins*)) **then** pc (cadr (*ins*), *program*)
    else cadr (*ins*) **endif**)

DEFINITION:
icode-push-ctrl-stk-free-size (*ins*, *pcn*, *program*)

=

```
' ((move_x_*)
  (sys-addr (full-ctrl-stk-addr . 0))
  (move_x_<x{s}>)
  (tpush_csp)
  (sub_<tsp>{s}_x{s}))
```


DEFINITION:
icode-push-global (*ins*, *pcn*, *program*)

=

list(`' (move_x_*)`,
    tag (`'addr`, cons (cadr (*ins*), **0**)),
    `' (tpush_<x{a}>))`

DEFINITION:
icode-push-local (*ins*, *pcn*, *program*)

=

list(`' (move_x_*)`,
    tag (`'nat`, offset-from-csp (cadr (*ins*), *program*)),
    `' (add_x{n}_csp)`,
    `' (tpush_<x{s}>))`

DEFINITION:
icode-push-temp-stk-free-size (*ins*, *pcn*, *program*)

=

```
' ((move_x_*)
  (sys-addr (full-temp-stk-addr . 0))
  (move_x_<x{s}>)
  (tpush_tsp)
  (sub_<tsp>{s}_x{s}))
```

DEFINITION:
icode-push-temp-stk-index (*ins*, *pcn*, *program*)
=
```
list('(move_y_tsp),
    '(move_x_*),
    '(sys-addr (empty-temp-stk-addr . 0)),
    '(move_x_<x{s}>),
    '(sub_<z>_x{s}_y{s}),
    '(tpush_x),
    '(move_x_*),
    tag('nat, 1+ cadr(ins)),
    '(sub_<tsp>{n}_x{n}))
```

DEFINITION:
icode-pushj (*ins*, *pcn*, *program*)
=
```
list('(tpush_*),
    tag('pc, cons(name(program), 1+ pcn)),
    '(jump_*),
    pc(cadr(ins), program))
```

DEFINITION:
icode-ret (*ins*, *pcn*, *program*)
=
```
list('(jump_*),
    tag('pc, cons(name(program), length(program-body(program)))))
```

DEFINITION:
icode-rsh-bitv (*ins*, *pcn*, *program*)
=
```
'((lsr_<tsp>_<tsp>{v}))
```

DEFINITION:
icode-set-global (*ins*, *pcn*, *program*)
=
```
list('(move_x_*),
    tag('addr, cons(cadr(ins), 0)),
    '(move_<x{a}>_<tsp>))
```

DEFINITION:
icode-set-local (*ins*, *pcn*, *program*)
=
```
list('(move_x_*),
    tag('nat, offset-from-csp(cadr(ins), program)),
    '(add_x{n}_csp),
    '(move_<x{s}>_<tsp>))
```

DEFINITION:
icode-sub-addr (*ins*, *pcn*, *program*)
=

```
'((tpop_x) (sub_<tsp>{a}_x{n}))
```

DEFINITION:
icode-sub-int (*ins*, *pcn*, *program*)
=

```
'((tpop_x) (sub_<tsp>{i}_x{i}))
```

DEFINITION:
icode-sub-int-with-carry (*ins*, *pcn*, *program*)
=

```
'((tpop_y)
  (tpop_x)
  (asr_<c>_<tsp>_<tsp>{b})
  (subb_<v>_x{i}_y{i})
  (move-v_<tsp>_*)
  (bool t)
  (tpush_x))
```

DEFINITION:
icode-sub-nat (*ins*, *pcn*, *program*)
=

```
'((tpop_x) (sub_<tsp>{n}_x{n}))
```

DEFINITION:
icode-sub-nat-with-carry (*ins*, *pcn*, *program*)
=

```
'((tpop_y)
  (tpop_x)
  (asr_<c>_<tsp>_<tsp>{b})
  (subb_<c>_x{n}_y{n})
  (move-c_<tsp>_*)
  (bool t)
  (tpush_x))
```

DEFINITION:
icode-sub1-int (*ins*, *pcn*, *program*)
=

```
'((decr_<tsp>_<tsp>{i}))
```

DEFINITION:
icode-sub1-nat (*ins*, *pcn*, *program*)
=

```
'((decr_<tsp>_<tsp>{n}))
```

DEFINITION:
icode-test-bitv-and-jump (*ins, pcn, program*)
=
if cadr (*ins*) = `'all-zero`
 then list (`'(tpop{v}_<z>_y)`,
      `'(move_x_*)`,
       pc (caddr (*ins*), *program*),
      `'(jump-z_x)`)
 else list (`'(tpop{v}_<z>_y)`,
      `'(move_x_*)`,
       pc (caddr (*ins*), *program*),
      `'(jump-nz_x)`) **endif**

DEFINITION:
icode-test-bool-and-jump (*ins, pcn, program*)
=
if cadr (*ins*) = `'t`
 then list (`'(tpop{b}_<z>_y)`,
      `'(move_x_*)`,
       pc (caddr (*ins*), *program*),
      `'(jump-nz_x)`)
 else list (`'(tpop{b}_<z>_y)`,
      `'(move_x_*)`,
       pc (caddr (*ins*), *program*),
      `'(jump-z_x)`) **endif**

DEFINITION:
icode-test-int-and-jump (*ins*, *pcn*, *program*)
=
case on cadr (*ins*):
  case = `zero`
   then list(' `(tpop{i}_<zn>_y)`,
         ' `(move_x_*)`,
         pc (caddr (*ins*), *program*),
         ' `(jump-z_x)`)
  case = `not-zero`
   then list(' `(tpop{i}_<zn>_y)`,
         ' `(move_x_*)`,
         pc (caddr (*ins*), *program*),
         ' `(jump-nz_x)`)
  case = `neg`
   then list(' `(tpop{i}_<zn>_y)`,
         ' `(move_x_*)`,
         pc (caddr (*ins*), *program*),
         ' `(jump-n_x)`)
  case = `not-neg`
   then list(' `(tpop{i}_<zn>_y)`,
         ' `(move_x_*)`,
         pc (caddr (*ins*), *program*),
         ' `(jump-nn_x)`)
  case = `pos`
   then list(' `(tpop{i}_<zn>_y)`,
         ' `(move_x_*)`,
         tag (' `pc`, cons (name (*program*), $1+ pcn$)),
         ' `(jump-n_x)`,
         ' `(jump-z_x)`,
         ' `(jump_*)`,
         pc (caddr (*ins*), *program*))
  otherwise list(' `(tpop{i}_<zn>_y)`,
            ' `(move_x_*)`,
            pc (caddr (*ins*), *program*),
            ' `(jump-n_x)`,
            ' `(jump-z_x)`) **endcase**

DEFINITION:
icode-test-nat-and-jump (*ins, pcn, program*)
=

if cadr (*ins*) = ′ **zero**
 then list (′ **(tpop{n}_<z>_y)**,
            ′ **(move_x_*)**,
            pc (caddr (*ins*), *program*),
            ′ **(jump-z_x)**)
 else list (′ **(tpop{n}_<z>_y)**,
            ′ **(move_x_*)**,
            pc (caddr (*ins*), *program*),
            ′ **(jump-nz_x)**) endif

DEFINITION:
icode-xor-bitv (*ins, pcn, program*)
=

′ **((tpop_x) (xor_<tsp>{v}_x{v}))**

DEFINITION:
icode1 (*ins, pcn, prog*)
=

case on car (*ins*):
 case = **call**
 then icode-call (*ins, pcn, prog*)
 case = **ret**
 then icode-ret (*ins, pcn, prog*)
 case = **locn**
 then icode-locn (*ins, pcn, prog*)
 case = **push-constant**
 then icode-push-constant (*ins, pcn, prog*)
 case = **push-local**
 then icode-push-local (*ins, pcn, prog*)
 case = **push-global**
 then icode-push-global (*ins, pcn, prog*)
 case = **push-ctrl-stk-free-size**
 then icode-push-ctrl-stk-free-size (*ins, pcn, prog*)
 case = **push-temp-stk-free-size**
 then icode-push-temp-stk-free-size (*ins, pcn, prog*)
 case = **push-temp-stk-index**
 then icode-push-temp-stk-index (*ins, pcn, prog*)
 case = **jump-if-temp-stk-full**
 then icode-jump-if-temp-stk-full (*ins, pcn, prog*)
 case = **jump-if-temp-stk-empty**
 then icode-jump-if-temp-stk-empty (*ins, pcn, prog*)
 case = **pop**
 then icode-pop (*ins, pcn, prog*)
 case = **pop***
 then icode-pop* (*ins, pcn, prog*)
 case = **popn**

then icode-popn (*ins, pcn, prog*)
case = **pop-local**
then icode-pop-local (*ins, pcn, prog*)
case = **pop-global**
then icode-pop-global (*ins, pcn, prog*)
case = **pop-locn**
then icode-pop-locn (*ins, pcn, prog*)
case = **pop-call**
then icode-pop-call (*ins, pcn, prog*)
case = **fetch-temp-stk**
then icode-fetch-temp-stk (*ins, pcn, prog*)
case = **deposit-temp-stk**
then icode-deposit-temp-stk (*ins, pcn, prog*)
case = **jump**
then icode-jump (*ins, pcn, prog*)
case = **jump-case**
then icode-jump-case (*ins, pcn, prog*)
case = **pushj**
then icode-pushj (*ins, pcn, prog*)
case = **popj**
then icode-popj (*ins, pcn, prog*)
case = **set-local**
then icode-set-local (*ins, pcn, prog*)
case = **set-global**
then icode-set-global (*ins, pcn, prog*)
case = **test-nat-and-jump**
then icode-test-nat-and-jump (*ins, pcn, prog*)
case = **test-int-and-jump**
then icode-test-int-and-jump (*ins, pcn, prog*)
case = **test-bool-and-jump**
then icode-test-bool-and-jump (*ins, pcn, prog*)
case = **test-bitv-and-jump**
then icode-test-bitv-and-jump (*ins, pcn, prog*)
case = **no-op**
then icode-no-op (*ins, pcn, prog*)
case = **add-addr**
then icode-add-addr (*ins, pcn, prog*)
case = **sub-addr**
then icode-sub-addr (*ins, pcn, prog*)
case = **eq**
then icode-eq (*ins, pcn, prog*)
case = **lt-addr**
then icode-lt-addr (*ins, pcn, prog*)
case = **fetch**
then icode-fetch (*ins, pcn, prog*)
case = **deposit**
then icode-deposit (*ins, pcn, prog*)
case = **add-int**
then icode-add-int (*ins, pcn, prog*)
case = **add-int-with-carry**

then icode-add-int-with-carry (*ins, pcn, prog*)
case = add1-int
then icode-add1-int (*ins, pcn, prog*)
case = sub-int
then icode-sub-int (*ins, pcn, prog*)
case = sub-int-with-carry
then icode-sub-int-with-carry (*ins, pcn, prog*)
case = sub1-int
then icode-sub1-int (*ins, pcn, prog*)
case = neg-int
then icode-neg-int (*ins, pcn, prog*)
case = lt-int
then icode-lt-int (*ins, pcn, prog*)
case = int-to-nat
then icode-int-to-nat (*ins, pcn, prog*)
case = add-nat
then icode-add-nat (*ins, pcn, prog*)
case = add-nat-with-carry
then icode-add-nat-with-carry (*ins, pcn, prog*)
case = add1-nat
then icode-add1-nat (*ins, pcn, prog*)
case = sub-nat
then icode-sub-nat (*ins, pcn, prog*)
case = sub-nat-with-carry
then icode-sub-nat-with-carry (*ins, pcn, prog*)
case = sub1-nat
then icode-sub1-nat (*ins, pcn, prog*)
case = lt-nat
then icode-lt-nat (*ins, pcn, prog*)
case = mult2-nat
then icode-mult2-nat (*ins, pcn, prog*)
case = mult2-nat-with-carry-out
then icode-mult2-nat-with-carry-out (*ins, pcn, prog*)
case = div2-nat
then icode-div2-nat (*ins, pcn, prog*)
case = or-bitv
then icode-or-bitv (*ins, pcn, prog*)
case = and-bitv
then icode-and-bitv (*ins, pcn, prog*)
case = not-bitv
then icode-not-bitv (*ins, pcn, prog*)
case = xor-bitv
then icode-xor-bitv (*ins, pcn, prog*)
case = rsh-bitv
then icode-rsh-bitv (*ins, pcn, prog*)
case = lsh-bitv
then icode-lsh-bitv (*ins, pcn, prog*)
case = or-bool
then icode-or-bool (*ins, pcn, prog*)
case = and-bool

then icode-and-bool (*ins, pcn, prog*)
case = **not-bool**
then icode-not-bool (*ins, pcn, prog*)
otherwise ' ((**error**))   **endcase**

DEFINITION:
icompile (*programs*)

=

if nlistp (*programs*) **then nil**
 else cons (icompile-program (car (*programs*)), icompile (cdr (*programs*))) **endif**

DEFINITION:
icompile-program (*program*)

=

cons (name (*program*),
      append (generate-prelude (*program*),
              append (icompile-program-body (program-body (*program*),
                                             0, *program*),
                      generate-postlude (*program*)))))

DEFINITION:
icompile-program-body (*lst, pcn, program*)

=

if nlistp (*lst*) **then nil**
 else append (icode (car (*lst*), *pcn, program*),
              icompile-program-body (cdr (*lst*), 1+ *pcn, program*)) **endif**

DEFINITION:
int-to-v (*i, l*)

=

if ilessp (*i*, 0) **then** nat-to-v (iplus (*i*, 2^l), *l*)
 else nat-to-v (*i, l*) **endif**

DEFINITION:
ipc-to-v (*pcpp, prog-links, word-size*)

=

nat-to-v (absolute-address (*pcpp, prog-links*), *word-size*)

DEFINITION:
jump_*-lst (*lst, program*)

=

if nlistp (*lst*) **then nil**
 else cons (' (**jump_***),
              cons (pc (car (*lst*), *program*), jump_*-lst (cdr (*lst*), *program*))) **endif**

DEFINITION:
label-address (*label*, *prog-label-tables*)
=

base-address (*label*, label-links (*label*, *prog-label-tables*))

DEFINITION:
label-links (*label*, *prog-label-tables*)
=

cdr (assoc (adp-name (*label*), *prog-label-tables*))

DEFINITION:
label-to-v (*ilab*, *prog-label-tables*, *word-size*)
=

nat-to-v (label-address (*ilab*, *prog-label-tables*), *word-size*)

DEFINITION:
link-area (*lst*, *link-tables*, *word-size*)
=

if nlistp (*lst*) **then nil**
 else cons (link-word (unlabel (car (*lst*)), *link-tables*, *word-size*),
          link-area (cdr (*lst*), *link-tables*, *word-size*)) **endif**

DEFINITION:
link-data-word (*x*, *link-tables*, *word-size*)
=

case on type (*x*):
 case = **nat then** nat-to-v (untag (*x*), *word-size*)
 case = **int then** int-to-v (untag (*x*), *word-size*)
 case = **bitv then** bitv-to-v (untag (*x*), *word-size*)
 case = **bool then** bool-to-v (untag (*x*), *word-size*)
 case = **addr then** addr-to-v (untag (*x*), usr-data-links (*link-tables*), *word-size*)
 case = **subr then** subr-to-v (untag (*x*), prog-links (*link-tables*), *word-size*)
 case = **sys-addr**
  then sys-addr-to-v (untag (*x*), sys-data-links (*link-tables*), *word-size*)
 case = **pc then** label-to-v (untag (*x*), prog-label-tables (*link-tables*), *word-size*)
 case = **ipc then** ipc-to-v (untag (*x*), prog-links (*link-tables*), *word-size*)
  otherwise nat-to-v (**0**, *word-size*) **endcase**

DEFINITION:
link-instr-word (*ins*, *word-size*)
=

mci (cadr (assoc (car (*ins*), link-instruction-alist)), *word-size*)

DEFINITION:
link-instruction-alist
=

```
'((add_<c>_x_x{n} (add (c) x x))
```

```
(add_<tsp>_<tsp>{v} (add nil (tsp) (tsp)))
(add_<tsp>_<tsp>{n} (add nil (tsp) (tsp)))
(add_<tsp>{a}_x{n} (add nil (tsp) x))
(add_tsp_*{n} (add nil tsp (pc 1)))
(add_tsp_x{n} (add nil tsp x))
(add_<tsp>{i}_x{i} (add nil (tsp) x))
(add_<tsp>{n}_x{n} (add nil (tsp) x))
(add_pc_x{n} (add nil pc x))
(add_x_x{n} (add nil x x))
(add_x{n}_csp (add nil x csp))
(addc_<c>_x{n}_y{n} (addc (c) x y))
(addc_<v>_x{i}_y{i} (addc (v) x y))
(and_<tsp>{v}_x{v} (and nil (tsp) x))
(and_<tsp>{b}_x{b} (and nil (tsp) x))
(asr_<c>_<tsp>_<tsp>{b} (asr (c) (tsp) (tsp)))
(cpop_cfp (move nil cfp (csp 1)))
(cpop_pc (move nil pc (csp 1)))
(cpush_* (move nil (-1 csp) (pc 1)))
(cpush_<tsp>+ (move nil (-1 csp) (tsp 1)))
(cpush_cfp (move nil (-1 csp) cfp))
(decr_<tsp>_<tsp>{i} (decr nil (tsp) (tsp)))
(decr_<tsp>_<tsp>{n} (decr nil (tsp) (tsp)))
(incr_<tsp>_<tsp>{i} (incr nil (tsp) (tsp)))
(incr_<tsp>_<tsp>{n} (incr nil (tsp) (tsp)))
(incr_y_y{n} (incr nil y y))
(int-to-nat (move nil x x))
(jump-n_x (move-n nil pc x))
(jump-nn_x (move-nn nil pc x))
(jump-nz_x (move-nz nil pc x))
(jump-z_x (move-z nil pc x))
(jump_* (move nil pc (pc)))
(jump_x{subr} (move nil pc x))
(lsr_<c>_x_x{n} (lsr (c) x x))
(lsr_<tsp>_<tsp>{v} (lsr nil (tsp) (tsp)))
(move-c_<tsp>_* (move-c nil (tsp) (pc 1)))
(move-v_<tsp>_* (move-v nil (tsp) (pc 1)))
(move-z_<tsp>_* (move-z nil (tsp) (pc 1)))
(move-n_x_* (move-n nil x (pc 1)))
(move_<tsp>_* (move nil (tsp) (pc 1)))
(move_<x{a}>_<tsp> (move nil (x) (tsp)))
(move_<x{s}>_<tsp> (move nil (x) (tsp)))
(move_cfp_csp (move nil cfp csp))
(move_csp_cfp (move nil csp cfp))
(move_x_* (move nil x (pc 1)))
(move_x_<x{s}> (move nil x (x)))
(move_x_tsp (move nil x tsp))
(move_x_x (move nil x x))
(move_y_* (move nil y (pc 1)))
(move_y_<y{s}> (move nil y (y)))
(move_y_tsp (move nil y tsp))
```

```
(neg_<tsp>_<tsp>{i} (neg nil (tsp) (tsp)))
(not_<tsp>_<tsp>{v} (not nil (tsp) (tsp)))
(or_<tsp>{v}_x{v} (or nil (tsp) x))
(or_<tsp>{b}_x{b} (or nil (tsp) x))
(sub_<c>_<tsp>{a}_x{a} (sub (c) (tsp) x))
(sub_<c>_<tsp>{n}_x{n} (sub (c) (tsp) x))
(sub_<nv>_<tsp>{i}_x{i} (sub (n v) (tsp) x))
(sub_<tsp>{a}_x{n} (sub nil (tsp) x))
(sub_x{s}_y{n} (sub nil x y))
(sub_<tsp>{i}_x{i} (sub nil (tsp) x))
(sub_<tsp>{n}_x{n} (sub nil (tsp) x))
(sub_<tsp>{s}_x{s} (sub nil (tsp) x))
(sub_<z>_x{s}_y{s} (sub (z) x y))
(subb_<c>_x{n}_y{n} (subb (c) x y))
(subb_<v>_x{i}_y{i} (subb (v) x y))
(tpop_<c>_x (move (c) x (tsp 1)))
(tpop_<x{a}> (move nil (x) (tsp 1)))
(tpop_<x{s}> (move nil (x) (tsp 1)))
(tpop_pc (move nil pc (tsp 1)))
(tpop_x (move nil x (tsp 1)))
(tpop_y (move nil y (tsp 1)))
(tpop{v}_<z>_y (move (z) y (tsp 1)))
(tpop{b}_<z>_y (move (z) y (tsp 1)))
(tpop{i}_<zn>_y (move (z n) y (tsp 1)))
(tpop{n}_<z>_y (move (z) y (tsp 1)))
(tpush_* (move nil (-1 tsp) (pc 1)))
(tpush_<x{a}> (move nil (-1 tsp) (x)))
(tpush_<x{s}> (move nil (-1 tsp) (x)))
(tpush_csp (move nil (-1 tsp) csp))
(tpush_tsp (move nil (-1 tsp) tsp))
(tpush_x (move nil (-1 tsp) x))
(xor_<tsp>_<tsp> (xor nil (tsp) (tsp)))
(xor_<tsp>{v}_x{v} (xor nil (tsp) x))
(xor_<tsp>{b}_*{b} (xor nil (tsp) (pc 1)))
(xor_<tsp>{b}_x{b} (xor nil (tsp) x))
(xor_<z>_<tsp>_x (xor (z) (tsp) x)))
```

DEFINITION:
link-mem (*boot-lst*,
        load-addr,
        usr-data-segment,
        prog-segment,
        sys-data-segment,
        link-tables,
        word-size)

=

append (boot-code (*boot-lst*, *load-addr*, *word-size*),
 append (link-segment (*usr-data-segment*, *link-tables*, *word-size*),
  append (link-segment (*prog-segment*, *link-tables*, *word-size*),
        link-segment (*sys-data-segment*, *link-tables*, *word-size*)))))

DEFINITION:
link-segment (*segment*, *link-tables*, *word-size*)

=

if nlistp (*segment*) **then nil**
 else append (link-area (cdar (*segment*), *link-tables*, *word-size*),
          link-segment (cdr (*segment*), *link-tables*, *word-size*)) **endif**

DEFINITION:
link-table-for-labels (*lst*, *addr0*)

=

if nlistp (*lst*) **then nil**
 elseif labeledp (car (*lst*))
 then cons (cons (cadar (*lst*), *addr0*), link-table-for-labels (cdr (*lst*), 1+ *addr0*))
 else link-table-for-labels (cdr (*lst*), 1+ *addr0*) **endif**

DEFINITION:
link-table-for-prog-labels (*segment*, *addr0*)

=

if nlistp (*segment*) **then nil**
 else cons (cons (caar (*segment*), link-table-for-labels (cdar (*segment*), *addr0*)),
        link-table-for-prog-labels (cdr (*segment*),
                         addr0 + length (cdar (*segment*))))) **endif**

DEFINITION:
link-table-for-segment (*segment*, *addr0*)

=

if nlistp (*segment*) **then nil**
 else cons (cons (caar (*segment*), *addr0*),
        link-table-for-segment (cdr (*segment*),
                         addr0 + length (cdar (*segment*))))) **endif**

DEFINITION:
link-word (*x*, *link-tables*, *word-size*)

=

if icode-instructionp (*x*) **then** link-instr-word (*x*, *word-size*)
 else link-data-word (*x*, *link-tables*, *word-size*) **endif**

DEFINITION:
load (*p*, *boot-lst*, *load-addr*)

=

m⇒fm9001 (i⇒m (r⇒i (p⇒r (*p*)), *boot-lst*, *load-addr*))

DEFINITION:
m⇒fm9001 (*m*)

=

list (list (ram-tree (m-regs (*m*), **16**),
        list (m-z-flg (*m*), m-n-flg (*m*), m-v-flg (*m*), m-c-flg (*m*))),
    ram-tree (m-mem (*m*), 2^{32}))

SHELL DEFINITION:
Add the shell 'm-state' of 6 arguments, with
recognizer function symbol 'm-statep', and
accessors 'm-regs',
            'm-c-flg', 'm-v-flg', 'm-n-flg', 'm-z-flg',
            'm-mem'.

DEFINITION:
mci (*ins*, *word-size*)

=

pack-instruction (extract-op (car (*ins*)),
                    extract-move-bits (car (*ins*)),
                    extract-cvnz (cadr (*ins*)),
                    extract-mode (caddr (*ins*)),
                    extract-reg (caddr (*ins*)),
                    extract-mode (cadddr (*ins*)),
                    extract-reg (cadddr (*ins*)),
                    word-size)

DEFINITION:
nat-0s (*n*)

=

if $n \cong 0$ **then nil**
 else cons (tag (**'nat**, **0**), nat-0s (*n* −1)) **endif**

DEFINITION:
offset-from-csp (*var*, *program*)

=

find-position-of-var (*var*, local-vars (*program*))

DEFINITION:
p⇒r (*p*)
=
r-state (p-pc (*p*),
          p⇒r_cfp (p-ctrl-stk (*p*), p-max-ctrl-stk-size (*p*)),
          p⇒r_csp (p-ctrl-stk (*p*), p-max-ctrl-stk-size (*p*)),
          p⇒r_tsp (p-temp-stk (*p*), p-max-temp-stk-size (*p*)),
          '(nat 0),
          '(nat 0),
          '(bool f),
          '(bool f),
          '(bool f),
          '(bool f),
          p-prog-segment (*p*),
          p-data-segment (*p*),
          p⇒r_sys-data-segment (p-ctrl-stk (*p*),
                                     p-max-ctrl-stk-size (*p*),
                                     p-temp-stk (*p*),
                                     p-max-temp-stk-size (*p*)),
          p-word-size (*p*),
          p-psw (*p*))

DEFINITION:
p⇒r_cfp (*stk*, *max*) = sub-addr (p⇒r_csp (pop (*stk*), *max*), 2)

DEFINITION:
p⇒r_csp (*stk*, *max*)
=
tag ('sys-addr, cons ('cstk, *max* − p-ctrl-stk-size (*stk*)))

DEFINITION:
p⇒r_ctrl-stk (*stk*, *max*)
=
cons ('cstk,
      append (nat-0s (*max* − p-ctrl-stk-size (*stk*)),
                append (p⇒r_ctrl-stk1 (*stk*, *max*), list (tag ('nat, 0)))))

DEFINITION:
p⇒r_ctrl-stk1 (*stk*, *max*)
=
if nlistp (*stk*) **then nil**
 else append (p⇒r_p-frame (top (*stk*), pop (*stk*), *max*),
                p⇒r_ctrl-stk1 (pop (*stk*), *max*)) **endif**

DEFINITION:
p⇒r_p-frame (*pframe*, *stk*, *max*)
=
append (strip-cdrs (bindings (*pframe*)), list (p⇒r_cfp (*stk*, *max*), ret-pc (*pframe*)))

DEFINITION:
p⟹r_sys-data-segment (*ctrl-stk, max-ctrl-stk-size, temp-stk, max-temp-stk-size*)
=
list (p⟹r_ctrl-stk (*ctrl-stk, max-ctrl-stk-size*),
    p⟹r_temp-stk (*temp-stk, max-temp-stk-size*),
    list (**'full-ctrl-stk-addr**,
        tag (**'sys-addr, '(cstk . 0)**))),
    list (**'full-temp-stk-addr**,
        tag (**'sys-addr, '(tstk . 0)**))),
    list (**'empty-temp-stk-addr**,
        tag (**'sys-addr**, cons (**'tstk**, *max-temp-stk-size*)))))

DEFINITION:
p⟹r_temp-stk (*temp-stk, max-temp-stk-size*)
=
cons (**'tstk**,
    append (nat-0s (*max-temp-stk-size* − length (*temp-stk*)),
        append (*temp-stk*, list (tag (**'nat, 0**)))))

DEFINITION:
p⟹r_tsp (*stk, max*)
=
tag (**'sys-addr**, cons (**'tstk**, *max* − length (*stk*)))

DEFINITION:
pack-instruction (*op, move-bits, cvnz, mode-b, reg-b, mode-a, reg-a, word-size*)
=
nat-to-v (($op \times 2^{24}$)
    +   (*move-bits* $\times 2^{20}$)
    +   (*cvnz* $\times 2^{16}$)
    +   (*mode-b* $\times 2^{14}$)
    +   (*reg-b* $\times 2^{10}$)
    +   (*mode-a* $\times 2^{4}$)
    +   *reg-a*,
    word-size)

DEFINITION:
prog-label-tables (*link-tables*) = cadr (*link-tables*)

DEFINITION:
prog-links (*link-tables*) = car (*link-tables*)

DEFINITION:
r⇒i (r)
=
i-state (r⇒i_pc (r-pc (r), r-prog-segment (r)),
        r-cfp (r),
        r-csp (r),
        r-tsp (r),
        r-x (r),
        r-y (r),
        r-c-flg (r),
        r-v-flg (r),
        r-n-flg (r),
        r-z-flg (r),
        icompile (r-prog-segment (r)),
        r-usr-data-segment (r),
        r-sys-data-segment (r),
        r-word-size (r),
        r⇒i_psw (r-psw (r)))

DEFINITION:
r⇒i_pc (*pc*, *programs*)
=
tag ('**ipc**,
    cons (area-name (*pc*),
            find-label (untag (*pc*),
                        cdr (icompile-program (definition (area-name (*pc*),
                                                           programs)))))))

DEFINITION:
r⇒i_psw (*psw*)
=
if *psw* = '**halt** **then** '**run**
 else *psw* **endif**

SHELL DEFINITION:
Add the shell 'r-state' of 15 arguments, with
recognizer function symbol 'r-statep', and
accessors 'r-pc', 'r-cfp', 'r-csp', 'r-tsp', 'r-x', 'r-y',
        'r-c-flg', 'r-v-flg', 'r-n-flg', 'r-z-flg',
        'r-prog-segment', 'r-usr-data-segment', 'r-sys-data-segment',
        'r-word-size', and 'r-psw'.

DEFINITION:
ram-tree (*lst*, *size*)

=

if *size* ≅ **0** **then** stub (nat-to-v (**0**, **32**))
  elseif nlistp (*lst*) **then** stub (nat-to-v (**0**, **32**))
  elseif *size* = **1** **then** ram (car (*lst*))
  else cons (ram-tree (firstn (*size* / **2**, *lst*), *size* / **2**),
        ram-tree (restn (*size* / **2**, *lst*), *size* / **2**)) **endif**

DEFINITION:
restn (*n*, *l*)

=

if listp (*l*)
 then if *n* ≅ **0** **then** *l*
      else restn (*n* −1, cdr (*l*)) **endif**
 else *l* **endif**

DEFINITION:
segment-length (*segment*)

=

if nlistp (*segment*) **then 0**
  else length (cdar (*segment*)) + segment-length (cdr (*segment*)) **endif**

DEFINITION:
strip-cdrs (*alist*)

=

if nlistp (*alist*) **then nil**
  else cons (cdar (*alist*), strip-cdrs (cdr (*alist*))) **endif**

DEFINITION:
subr-to-v (*subr, prog-links, word-size*)

=

nat-to-v (base-address (*subr, prog-links*), *word-size*)

DEFINITION:
sys-addr-to-v (*adp, sys-data-links, word-size*)

=

nat-to-v (absolute-address (*adp, sys-data-links*), *word-size*)

DEFINITION:
sys-data-links (*link-tables*) = cadddr (*link-tables*)

DEFINITION:
usr-data-links (*link-tables*) = caddr (*link-tables*)

Appendix V

The Formal
Correctness Theorem

The correctness theorem for the FM9001 implementation of Piton is

THEOREM: FM9001 Piton is Correct
$$
\begin{aligned}
(\quad & \text{proper-p-statep}\,(p_0) \\
\wedge\ & (load\text{-}addr \in \mathbf{N}) \\
\wedge\ & \text{p-loadablep}\,(p_0,\ load\text{-}addr) \\
\wedge\ & (\text{p-word-size}\,(p_0) = \mathbf{32}) \\
\wedge\ & (p_n = \text{p}\,(p_0,\ n)) \\
\wedge\ & (\neg\ \text{errorp}\,(\text{p-psw}\,(p_n))) \\
\wedge\ & (ts = \text{type-specification}\,(\text{p-data-segment}\,(p_n)))) \\
\to\ (\quad & \text{p-data-segment}\,(p_n) \\
=\ & \text{display-fm9001-data-segment}\,(\text{fm9001}\,(\text{load}\,(p_0,\ boot\text{-}lst,\ load\text{-}addr), \\
& \qquad\qquad\qquad\qquad\qquad\qquad \text{fm9001-clock}\,(p_0,\ n)), \\
& \qquad\qquad ts, \\
& \qquad\qquad \text{link-tables}\,(p_0,\ load\text{-}addr)))
\end{aligned}
$$

We have already presented the formal definitions of 'proper-p-statep', 'p', 'fm9001', 'load' and the accessors for the 'p-state' shell, including 'p-word-size', 'p-psw', and 'p-data-segment'. The remaining nonprimitive functions used in the statement of correctness are 'p-loadablep', 'errorp', 'type-specification', 'display-fm9001-data-segment' 'link-tables', and 'fm9001-clock'. All but the last are defined in this appendix.

Recall our discussion of the correctness theorem and in particular of the clock expression, fm9001-clock$(p_0,\ n)$. We regard 'fm9001-clock' as a "witness function" for what is informally understood as existential quantification on the number of steps that FM9001 must be run to duplicate a Piton computation. We do not exhibit the definition of 'fm9001-clock' in this book because we think of it as a definition specific to our particular proof. Suffice it to say that 'fm9001-clock' can be constructively defined—indeed, it *is* constructively defined in our proof. Its definition is somewhat larger than that of 'p' itself and is structurally isomorphic. Fm9001-clock$(p_0,\ n)$ steps forward from p_0. On each iteration 'fm9001-clock' uses a function specific to our implementation to determine how many FM9001 machine instructions are executed for the current Piton instruction in the current environment.[16] 'Fm9001-clock' sums the instruction counts and iterates until either *n* steps are taken or the Piton computation halts normally or with an error. With the exception of the footnote on page 145 we do not discuss 'fm9001-clock' further in this document.

We have already discussed at some length the concepts involved in the correctness theorem. The rest of this appendix is an alphabetical listing of the formal definitions of the remaining functions used in the theorem. The "entry points" into the definitions are the functions 'p-loadablep', 'errorp', 'type-specification', 'display-fm9001-data-segment' and 'link-tables'. It is noteworthy that all of the definitions listed in the preceding Appendices II-IV, plus those here, are necessary *simply to state* the correctness result. The *proof* of the correctness result requires the definition of 'fm9001-clock' and hundreds of other functions.

DEFINITION:
area-type-specification (*area*)
=
cons (car (*area*), type-lst (cdr (*area*)))

DEFINITION:
assoc-cdrp (*n, alist*)
=
if nlistp (*alist*) **then f**
 elseif *n* = cdar (*alist*) **then t**
 else assoc-cdrp (*n*, cdr (*alist*)) **endif**

DEFINITION:
display-fm9001-array (*type-lst, n, fm-mem, link-tables*)
=
if nlistp (*type-lst*) **then nil**
 else cons (unlink-data-word (car (*type-lst*),
                                 read-mem (nat-to-v (*n*, **32**), *fm-mem*),
                                 link-tables),
          display-fm9001-array (cdr (*type-lst*),
                                1+ *n*,
                                fm-mem, link-tables)) **endif**

---

[16]Sometimes the number of machine code instructions varies according to which path is taken through the i-code generated.

DEFINITION:
display-fm9001-data-area (*area-type-spec*, *fm-mem*, *link-tables*)
=

cons (car (*area-type-spec*),
    display-fm9001-array (cdr (*area-type-spec*),
                    base-address (car (*area-type-spec*),
                              usr-data-links (*link-tables*)),
             fm-mem,
             link-tables))

DEFINITION:
display-fm9001-data-segment (*fm-state*, *type-spec*, *link-tables*)
=

display-fm9001-data-segment1 (*type-spec*, cadr (*fm-state*), *link-tables*)

DEFINITION:
display-fm9001-data-segment1 (*type-spec*, *fm-mem*, *link-tables*)
=

if nlistp (*type-spec*) **then nil**
else cons (display-fm9001-data-area (car (*type-spec*), *fm-mem*, *link-tables*),
        display-fm9001-data-segment1 (cdr (*type-spec*),
                        fm-mem, *link-tables*)) **endif**

DEFINITION:
errorp (*psw*) = (*psw* ≠ **'run**) ∧ (*psw* ≠ **'halt**)

DEFINITION:
find-containing-area-name (*n*, *link-table*)
=

if nlistp (*link-table*) **then 0**
 elseif nlistp (cdr (*link-table*)) **then** caar (*link-table*)
 elseif (*n* ≥ cdar (*link-table*)) ∧ (*n* < cdadr (*link-table*))
 then caar (*link-table*)
 else find-containing-area-name (*n*, cdr (*link-table*)) **endif**

DEFINITION:
find-containing-label-table (*n*, *label-tables*)
=

if nlistp (*label-tables*) **then f**
 elseif assoc-cdrp (*n*, cdar (*label-tables*)) **then** cdar (*label-tables*)
 else find-containing-label-table (*n*, cdr (*label-tables*)) **endif**

DEFINITION:
invert-absolute-address (*n*, *link-table*)
=

cons (find-containing-area-name (*n*, *link-table*),
    n − base-address (find-containing-area-name (*n*, *link-table*), *link-table*))

DEFINITION:
invert-base-address (n, *link-table*)

=

find-containing-area-name (n, *link-table*)

DEFINITION:
invert-label-address (n, *prog-label-tables*)

=

invert-base-address (n, find-containing-label-table (n, *prog-label-tables*))

DEFINITION:
link-tables (p, *load-addr*) = i-link-tables (r$\Rightarrow$i (p$\Rightarrow$r (p)), *load-addr*)

DEFINITION:
p-loadablep (p, *load-addr*)

=

total-p-system-size (p, *load-addr*) < $2^{\text{p-word-size}\,(p)}$

DEFINITION:
total-p-system-size (p, *load-addr*)

=

    load-addr
+   segment-length (p-data-segment (p))
+   segment-length (icompile (p-prog-segment (p)))
+   (1+ p-max-ctrl-stk-size (p))
+   (1+ p-max-temp-stk-size (p))
+   **3**

DEFINITION:
type-lst (*lst*)

=

if nlistp (*lst*) **then nil**
 else cons (type (car (*lst*)), type-lst (cdr (*lst*))) **endif**

DEFINITION:
type-specification (*segment*)

=

if nlistp (*segment*) **then nil**
 else cons (area-type-specification (car (*segment*)),
          type-specification (cdr (*segment*))) **endif**

DEFINITION:
unlink-data-word (*type*, *v*, *link-tables*)
=

case on *type*:
  case = **nat** **then** tag (**'nat**, v-to-nat (*v*))
  case = **int** **then** tag (**'int**, v-to-int (*v*))
  case = **bitv** **then** tag (**'bitv**, v-to-bitv (*v*))
  case = **bool** **then** tag (**'bool**, v-to-bool (*v*))
  case = **addr** **then** tag (**'addr**, v-to-addr (*v*, usr-data-links (*link-tables*)))
  case = **subr** **then** tag (**'subr**, v-to-subr (*v*, prog-links (*link-tables*)))
  case = **sys-addr**
   then tag (**'sys-addr**, v-to-sys-addr (*v*, sys-data-links (*link-tables*)))
  case = **pc** **then** tag (**'pc**, v-to-label (*v*, prog-label-tables (*link-tables*)))
  otherwise **'(unrecognized i-level type)** **endcase**

DEFINITION:
v-to-addr (*v*, *usr-data-links*)
=

invert-absolute-address (v-to-nat (*v*), *usr-data-links*)

DEFINITION:
v-to-bitv (*v*)
=

if nlistp (*v*) **then nil**
 else append (v-to-bitv (cdr (*v*)),
           list (**if** car (*v*) **then 1**
                else 0 **endif**)) **endif**

DEFINITION:
v-to-bool (*v*)
=

if car (*v*) **then 't**
 else 'f **endif**

DEFINITION:
v-to-int (*v*)
=

if v-to-nat (*v*) $< 2^{(\text{length}\,(v)\,-1)}$ **then** v-to-nat (*v*)
 else idifference (v-to-nat (*v*), $2^{\text{length}\,(v)}$) **endif**

DEFINITION:
v-to-label (*v*, *prog-label-tables*)
=

invert-label-address (v-to-nat (*v*), *prog-label-tables*)

DEFINITION:

v-to-nat (*v*)

=

if nlistp (*v*) **then** 0
 else   **if** car (*v*) **then** 1
             else 0 **endif**
      +   (**2** × v-to-nat (cdr (*v*))) **endif**

DEFINITION:

v-to-subr (*v*, *prog-links*)

=

invert-base-address (v-to-nat (*v*), *prog-links*)

DEFINITION:

v-to-sys-addr (*v*, *sys-data-links*)

=

invert-absolute-address (v-to-nat (*v*), *sys-data-links*)

Bibliography

1. R. Aubin. Strategies for Mechanizing Structural Induction. International Joint Conference on Artificial Intelligence, 1977.

2. W. Bevier. *A Verified Operating System Kernel*. Ph.D. Th., University of Texas at Austin, 1987.

3. W.R. Bevier, W.A. Hunt, J S. Moore, and W.D. Young. "Special Issue on System Verification". *Journal of Automated Reasoning 5*, 4 (1989), 409-530.

4. J. Bowen, M. Franzle, E.R. Olderog, and A.P. Ravn. Developing Correct Systems. Proceedings of the 5th Euromicro Workshop on Real-Time Systems, 1993.

5. R. S. Boyer and J S. Moore. *A Computational Logic Handbook*. Academic Press, New York, 1988.

6. R.S. Boyer and Y. Yu. Automated Correctness Proofs of Machine Code Programs for a Commercial Microprocessor. In *11th Conference on Automated Deduction, Lecture Notes in Computer Science*, D. Kapur, Ed., Springer-Verlag, 1992, pp. 416-430.

7. B.C. Brock and W.A. Hunt. The Formal Specification and Verification of the FM9001 Microprocessor. Tech. Rept. Technical Report 86, Computational Logic, Inc., 1717 W. Sixth Street, Suite 290, Austin, TX 78703, October, 1994.

8. R. Burstall. "Proving Properties of Programs by Structural Induction". *The Computer Journal 12*, 1 (1969), 41-48.

9. R.M. Burstall and P.J. Landin. Programs and Their Proofs: an Algebraic Approach. In *Machine Intellgience 4*, B. Meltzer and D. Michie, Ed., Edinburgh University Press, 1969, pp. 17-43.

10. R. Cartwright. *A Practical Formal Semantic Definition and Verification System for Typed LISP*. Ph.D. Th., Stanford University, 1976.

11. L.M. Chirica and D.F. Martin. "Toward Compiler Implementation Correctness Proofs". *ACM Transaction on Programming Languages and Systems 8*, 2 (1986), 185-214.

12. A. Cohn. High Level Proof in LCF. Proceedings of the Fifth Symposium on Automated Deduction, 1979.

13. A.J. Cohn. "The Notion of Proof in Hardware Verification". *Journal of Automated Reasoning 5* (1989), 127-138.

14. W.J. Cullyer. Implementing Safety Critical Systems: The Viper Microprocessor. In *VLSI Specification, Verification and Synthesis*, G. Birtwistle and P.A. Subrahmanyam, Ed., Kluwer Academic Publishers, 1988, pp. 1-25.

15. P. Curzon. "Deriving Correctness Properties of Compiled Code". *Formal Methods in System Design 3*, 1/2 (1993), 83-115.

16. A. Flatau. *A Verified Implementation of an Applicative Language with Dynamic Storage Allocation*. Ph.D. Th., University of Texas, 1992. Also available through Computational Logic, Inc., Suite 290, 1717 West Sixth Street, Austin, TX 78703.

17. M. Gordon. HOL: A Proof Generating System for Higher-Order Logic. Tech. Rept. 103, University of Cambridge, Computer Laboratory, 1987.

18. W.A. Hunt, Jr. *FM8501: A Verified Microprocessor*. LNAI Number 795, Springer-Verlag, 1994.

19. J.J. Joyce. A Verified Compiler for a Verified Microprocessor. Tech. Rept. 167, University of Cambridge, Computer Laboratory, 1989.

20. J.J. Joyce. Totally Verified Systems: Linking Verified Software to Verified Hardware. Specification, Verification and Synthesis: Mathematical Aspects, 1989.

21. M. Kaufmann. "An Extension of the Boyer-Moore Theorem Prover to Support First-Order Quantification". *J. Automated Reasoning 9*, 3 (December 1992), 355-372.

22. K. Kunen. A Ramsey Theorem in Boyer-Moore Logic. Tech. Rept. http://www.cs.wisc.edu/~kunen/kunen.html, Computer Sciences Department, University of Wisconsin, 1994.

23. R.L. London. Correctness of a Compiler for a LISP Subset. Proceedings of an ACM Conference on Proving Assertions about Programs, 1972.

24. J. McCarthy and J. Painter. Correctness of a Compiler for Arithmetic Expressions. Proceeding of Symposium on Applied Mathematics, American Mathematical Society, 1967.

25. R. Milner and R. Weyhrauch. Proving Compiler Correctness in a Mechanized Logic. In *Machine Intelligence 7*, Edinburgh University Press, Edinburgh, Scotland, 1972, pp. 51-70.

26. J S. Moore. Piton: A Verified Assembly-Level Language. Tech. Rept. CLI-22, Computational Logic, Inc., Austin, Tx, June, 1988.

27. F.L. Morris. Advice of Structuring Compilers and Proving Them Correct. Proceedings of the ACM Symposium on Principles of Programming Languages, October, 1973, pp. 144-152.

28. D.P. Oliva and M. Wand. A Verified Compiler for Pure PreScheme. Tech. Rept. NU-CCS-92-5, Northeastern University College of Computer Science, 1992.

29. W. Polak. *Compiler Specification and Verification.* Springer-Verlag, Berlin, 1981.

30. D.M. Russinoff. "A Mechanical Proof of Quadratic Reciprocity". *Journal of Automated Reasoning 8*, 1 (1992), 3-21.

31. N. Shankar. *Metamathematics, Machines, and Gödel's Proof.* Cambridge University Press, 1994.

32. J. R. Shoenfield. *Mathematical Logic.* Addison-Wesley, Reading, Ma., 1967.

33. M. Wilding. A Mechanically Verified Application for a Mechanically Verified Environment. Proceedings of CAV '93, LNCS 697, 1993, pp. 268-279.

34. W.D. Young. A Verified Code-Generator for a Subset of Gypsy. University of Texas at Austin, 1988. Also available through Computational Logic, Inc., Suite 290, 1717 West Sixth Street, Austin, TX 78703.

35. Y. Yu. *Automated Proofs of Object Code for a Widely Used Microprocessor.* Ph.D. Th., University of Texas, 1992.

Index

The numbers associated with each entry of this index are page numbers. Each number is in one of three fonts. Bold face numbers, such as **27** and **135**, indicate the defining occurrence of the symbol or phrase. Numbers in Roman font, such as 27 and 135, indicate significant occurrences of the symbol or phrase in text. Not every occurrence of the symbol or phrase in text is deemed "significant." Numbers in italic font, such as *27* and *135*, indicate occurrences of the given symbol in the definitions listed in Appendices II-V of this report. Every such occurrence is noted. Such page numbers indicate the beginning of the containing definition, rather than the page on which the occurrence is found.